U0164971

大学英语4级考试
710分全真模拟试卷及详解（修订本）

主 编：王 波　　张 琳　　周天楠

副主编：徐英辉　　历冬风　　周巍巍

　　　　赵 巍　　何丽萍　　韩 琦

　　　　唐俊莉

石油工业出版社

修订本及编写说明

1. 2006 年 9 月，我们组织编写并出版了《大学英语 4 级考试 710 分全真模拟试卷及详解》这本书，经过两年的市场验证，各方面特别是参加四六级考试的大学生反映很好，深受大学生读者的欢迎，他们认为该书最大的特点及优势就是**快速阅读及仔细阅读**部分题文呼应，直观明了，并有文章大意和详尽的解释说明，写作部分设有专家点评，并列出 4 级常用句型。该书销售一直很好，据统计，2006 年该书曾名列北京地区大学英语四、六级考试辅导类丛书的前几位。随着时间的推移，本书的许多内容急需较大的调整，这样才能更好地帮助考生理解新的命题趋势，准确把握新的题型变化，了解新的考查方向，这样学生才能在考试中取得优异成绩。鉴于此，我们针对市场需求和近几年大学英语四六级考试的新情况，紧扣最新考试大纲，顺应广大读者的要求，重新编辑出版《大学英语 4 级考试 710 分全真模拟试卷及详解》，其中，我们根据新的命题趋势、题型变化，以及新的考查方向，对第一套、第二套题进行了整体替换，同时对其余六套题进行了修订。

2. 本书是在对五种新题型要求和命题规律深入研究的基础上，对其相对应的解题思路进行了全面诠释，这些解题方法和技巧简单明了，实用有效且实战性强，其配套的专项训练和综合模拟题，能让考生在短时间内迅速熟悉新题型，进而攻克此类新题型。

3. 本书编者都是国内大学英语教学一线的专家和教授，他们是在充分理解国家教育部对大学英语教学新要求的基础上进行编写的，而且把这种精神全面、彻底地贯穿到本书各个部分的编写中。

4. 本书结合国际上主流英语测试的权威考题，精心编撰，试题严谨科学、难度适中、预测性强。

5. 讲练结合：本书的详解可谓是本书的一大特色，精心设计的练习可以帮助考生巩固、强化考试所需的技能，详解可以让学生真正理解习题，知道自己错在哪里，有方向性，有针对性。为了使广大考生更好地使用本书，现将本书特点图示如下。

Part Ⅰ Writing
写作设有专家点评，并列出 4 级常用句型。

It has long been taught in school textbook that man has the initiative and power to transform nature to man's advantage. Yet recently with the environment degenerating at unconceivable speed…

【评论】本文是一篇议论文，主要论述了人和自然之间的关系……

【常用句型】　第一段：the fact turns out to be that…The more…, the more…

Part Ⅱ Reading Comprehension (Skimming and Scanning)
快速阅读，分析文章框架，题文呼应。

Others were forced out of their homelands in order to escape atrocities (暴政，暴行).The two world wars during this century forced large numbers of people to seek refuge in a distant land in order to survive.

同义转述

2. World War Ⅱ was a key cause of the immigration boom in the past century. (Y)

Part Ⅲ Listening Comprehension

听力由美籍教师朗读，严格按4级考试要求录制。

11．M: How much are these jackets?

W: They are on sale today, sir. Twenty-five dollars each, or two for forty dollars.

Q: How much does one jacket cost?

【答案及解析】本题属于推理题。男士问："这些夹克衫多少钱？" 女士说："今天削价处理，25美元一件，40美元两件。"可推出正确答案是 A）Twenty-five dollars. 25美元一件。

Part Ⅳ Reading Comprehension (Reading in Depth)

仔细阅读题文呼应，直观明了，并有文章大意和详尽的解释说明。

Nor is that place irrelevant to the healthy development of the child. The family is a co-operative enterprise for which it is difficult to lay down rules. Because each family needs to work out its own ways for solving its own problems.

同义
转述

59. According to the author, the solution of family problems _____.

A) is best left in hands of women

B) is similar in all families

C) is not necessary in household where sharing is done

D) needs to be reached by ways unique to each family

59.【题目译文】根据作者的观点，家庭问题的解决方法是_____ 。

【答案及解析】D) 根据第三段最后一句话，每个家庭都需要找出解决自己问题的不同方法，由此可知 D)为正确答案。

本书由王波、张琳、周天楠任主编，王波编写字数为 8 万字，张琳、周天楠各编写 6 万字，其他副主编徐英辉、历冬风、周巍巍、赵巍、何丽萍、韩琦、唐俊莉各 5 万字。

目　　录

大学英语 4 级考试新题型模拟试卷一答案及详解 …………………………………………1

大学英语 4 级考试新题型模拟试卷二答案及详解 …………………………………………20

大学英语 4 级考试新题型模拟试卷三答案及详解 …………………………………………37

大学英语 4 级考试新题型模拟试卷四答案及详解 …………………………………………54

大学英语 4 级考试新题型模拟试卷五答案及详解 …………………………………………73

大学英语 4 级考试新题型模拟试卷六答案及详解 …………………………………………91

大学英语 4 级考试新题型模拟试卷七答案及详解 ………………………………………109

大学英语 4 级考试新题型模拟试卷八答案及详解 ………………………………………127

大学英语4级考试新题型模拟试卷一答案及详解

参考答案

Part I   Writing

The Benefits of Volunteering

Volunteering has grown into a fashion these years, with volunteers ranging from college students to white collar workers. It is reported now China has 4.5 million registered volunteers having provided more than 4.5 billion hours of volunteer work. Why would so many people like to do volunteer work? Are there any benefits of volunteering?

The answer is absolutely yes. First of all, through volunteering we can gain more experiences. Whatever volunteering work we do, either being a volunteer teacher for a village school or a volunteer worker in a retired house, we can taste different lifestyles and broaden our horizon.

Secondly, through volunteering we can acquire the sense of being needed. Many people feel frustrated with travails in life and empty with the cycling of working and resting. They gradually become numb, losing the sense of being needed. By joining the volunteer group and offering help voluntarily to people who need it, we may feel happy and fulfilled again.

Thirdly, through volunteering we can learn more good qualities. Volunteer work is team work, which needs good cooperation. We can know people with different personalities and different background joining in the volunteer group and devoting to the volunteer work together. From our co-volunteers we can absorb many good qualities that could help us be better persons.

In conclusion, volunteering is of great benefit in that we can gain more experiences, acquire the sense of being needed and learn more good qualities.

Part II   Reading Comprehension (Skimming and Scanning)

1. Y   2. N   3. Y   4. N   5. N   6. N   7. Y

8. a heat source

9. to its volume

10. wind

Part III   Listening Comprehension

Section A

11. A   12. D   13. D   14. B   15. A   16. D   17. D   18. C   19. D   20. B

21. B   22. C   23. A   24. A   25. D

Section B

26. C   27. D   28. D   29. C   30. C   31. B   32. D   33. B   34. D   35. D

Section C

36. destructive   37. seldom   38. sets   39. shallower

40. bottom   41. strength   42. shore   43. strike

44. strike the shore with a force of 75 million pounds

45. the waves, no matter how big or how violent, affect only the surface of the sea

46. the water a hundred fathoms (600 feet) beneath the surface is just as calm as on the day without a breath of wind

1

Part IV   Reading Comprehension (Reading in Depth)

Section A

47. I: happened   48. L: informal   49. M: interruption   50. J: definitely   51. K: physical

52. D: moral   53. A: honor   54. C: had   55. H: power   56. B: accomplish

Section B

57. B   58. A   59. B   60. D   61. C   62. B   63. C   64. C   65. B   66. A

Part V   Cloze

67. A   68. C   69. A   70. B   71. D   72. C   73. B   74. C   75. B   76. D

77. C   78. A   79. D   80. D   81. D   82. D   83. B   84. D   85. C   86. B

Part VI   Translation

87. regardless of   88. So far/ Up to now/ Until now

89. on time   90. instead of   91. on purpose

答案解析及录音原文

Part I   Writing

The Benefits of Volunteering

Volunteering has grown into a fashion these years, with volunteers ranging from college students to white collar workers. It is reported now China has 4.5 million registered volunteers having provided more than 4.5 billion hours of volunteer work. Why would so many people like to do volunteer work? Are there any benefits of volunteering?

The answer is absolutely yes. First of all, through volunteering we can gain more experiences. Whatever volunteering work we do, either being a volunteer teacher for a village school or a volunteer worker in a retired house, we can taste different lifestyles and broaden our horizon.

Secondly, through volunteering we can acquire the sense of being needed. Many people feel frustrated with travails in life and empty with the cycling of working and resting. They gradually become numb, losing the sense of being needed. By joining the volunteer group and offering help voluntarily to people who need it, we may feel happy and fulfilled again.

Thirdly, through volunteering we can learn more good qualities. Volunteer work is team work, which needs good cooperation. We can know people with different personalities and different background joining in the volunteer group and devoting to the volunteer work together. From our co-volunteers we can absorb many good qualities that could help us be better persons.

In conclusion, volunteering is of great benefit in that we can gain more experiences, acquire the sense of being needed and learn more good qualities.

【评论】本文第一段利用事实及相关数据引出"志愿者"这一话题，随即提出设问，引出下文。在接下来的第二、三、四段回答所提的问题，最后一段表明了自己的看法。本文论证条理清晰，语言丰富，分析周全。

【常用句型】

第一段：提出问题 Why would so many people like to do volunteer work? Are there any benefits of volunteering?

第二段：阐述观点 The answer is absolutely yes.

First of all, through volunteering we can gain more experiences.

Secondly, through volunteering we can acquire the sense of being needed.

Thirdly, through volunteering we can learn more good qualities.

第三段：点明自己的观点 In conclusion, volunteering is of great benefit in that …

Part Ⅱ　　　　Reading comprehension　　　　(Skimming and Scanning)

【文章及答案解析】文章主要讲的是 Wildfires，分四个小标题，分别是 Fire Starts (火引子)、Fueling the Flames (给火焰加燃料)、Wind and Rain (风和雨)、Fire on the Mountain (山上的火灾)可知本文讲的是野火如何产生和持续。

Wildfires

In just seconds, a spark or even the sun's heat alone sets off an extremely large fire. The wildfire quickly spreads, consuming the thick, dried-out plants and almost everything else in its path. What was once a forest becomes a virtual powder keg of untapped fuel? In a seemingly instantaneous burst, the wildfire overtakes thousands of acres of surrounding land, threatening the homes and lives of many in the vicinity.

Fire Starts

On a hot summer day, when drought conditions peak, something as small as a spark from a train car's wheel striking the track can ignite a raging wildfire. Sometimes, fires occur naturally, ignited by heat from the sun or a lightening strike. However, the majority of wildfires are the result of human carelessness.

Common causes for wildfires include:

Arson

Campfires

Discarding lit cigarettes

Improperly burning debris

Playing with matches or fireworks

Prescribed fires

Everything has a temperature at which it will burst into flames. This temperature is called a material's flash point. Wood's flash point is 572 degrees Fahrenheit (300℃). When wood is heated to this temperature, it releases hydrocarbon gases that mix with oxygen in the air, combust and create fire.

1. This passage explores how wildfires are born and live. (Y)

【文章大意】　　　　野火

仅仅只是短短的数秒时间，一点火光甚至只是太阳的热量就能变成地狱。森林大火蔓延得很快，烧毁了其蔓延途中稠密的干的植被和几乎其他的每一样东西。森林就变成了未开启的虚拟引火气罐。从表面上看是瞬间爆发，野火覆盖了周围几千英亩的陆地，威胁着周围的家园和生命。

火的始作俑者

在炎炎夏日的一天，当达到最干旱的条件时，像被汽车轮胎压过的如火光一样小的东西就能点燃一场森林大火。有时火是由太阳或闪电袭击而自然产生的。然而，绝大多数森林大火是由于人的疏忽造成的。

引起森林大火的原因包括：

同义转述

3. Wildfires are mainly caused by people, not by nature. (Y)

故意纵火

篝火

丢弃的燃着的香烟

不当燃烧的碎片

玩火柴或火焰

按规定生的火

细节确认

2. Wood will burst into flames at 300 degrees Fahrenheit. (N)

3

There are three components needed for ignition and combustion to occur. A fire requires fuel to burn, air to supply oxygen, and a heat source to bring the fuel up to ignition temperature. Heat, oxygen and fuel form the fire triangle. Firefighters often talk about the fire triangle when they are trying to put out a blaze. The idea is that if they can take away any one of the pillars of the triangle, they can control and ultimately extinguish the fire. After combustion occurs and a fire begins to burn, there are several factors that determine how the fire spreads. These three factors include fuel, weather and topography. Depending on these factors, a fire can quickly fizzle or turn into a raging blaze that scorches thousands of acres.

Fueling the Flames

Wildfires spread based on the type and quantity of fuel that surrounds it. Fuel can include everything from trees, underbrush and dry grassy field to homes. The amount of flammable material that surrounds a fire is referred to as the fuel load. Fuel load is measured by the amount of available fuel per unit area, usually tons per acre.

A small fuel load will cause a fire to burn and spread slowly, with a low intensity. If there are a lot of fuels, the fire will burn more intensely, causing it to spread faster. The faster it heats up the material around it, the faster those materials can ignite. The dryness of the fuel can also affect the behavior of the fire. When the fuel is very dry, it is consumed much faster and creates a fire that is much more difficult to contain.

Here are the basic fuel characteristics that decide how it affects a fire.

Size and shape

Arrangement

Moisture content

Small fuel materials, also called flashy fuels, such

8. There are three components needed for ignition and combustion to occur: fuel, air and _____. (**a heat source**)

每样东西都有使它燃烧的温度。这个温度就叫做物质的燃点。木头的燃点是 572 华氏度（300 摄氏度）。当木头加热到这个温度，就会释放氢气，和空气中的氧气混在一起燃烧起来。

要燃烧起来必须要有三种因素。一把火需要燃料来燃烧、空气来提供氧气，还有使燃料到达燃烧温度的热源。热量、氧气和燃料是火的三要素。消防员在灭火时常常提到这三要素。这种观念就是如果他们除去三要素中的一种，他们就能控制并最终灭掉火。

燃烧发生后，将开始一场大火时，有几个因素决定了火是怎样蔓延的。这三种因素包括燃料、天气和地形。依靠这些因素，火可以很快被熄灭或者大范围地燃起来并烧掉数千英亩的地区。

大火的燃料

森林大火是基于周围的燃料的类型和质量蔓延的。燃料包括从树木、丛林及干草地到家里的每一样东西。火周围的易燃物数量被视为可燃物负荷。可燃物负荷是测量每块地区可得燃料的数量，通常是每英亩的吨数。

4. In most situations, we measure fuel load by tons per hectare. (**N**)

小的可燃物负荷会引起一场火的燃烧并慢慢地低强度地蔓延。如果有很多燃料，火就会燃烧得很快，导致蔓延得更快。火周围的物质加热得越快，那些物质就越易点燃。燃料的干燥程度也会影响火的变化。燃料越干燥，它就会点燃得越快，产生的火就更难熄灭。

下面是基本的影响火势的燃料特性：

大小和形状

放置

潮湿程度

as dry grass, pine needles, dry leaves, twigs and other dead brush, burn faster than large logs or stumps (this is why you start a fire with kindling rather than logs). On a chemical level, different fuel materials take longer to ignite than others. But in a wildfire, where most of the fuel is made of the same sort of material, the main variable in ignition time is the ratio of the fuel's total surface area to its volume. Since a twig's surface area is much larger than its volume, it ignites quickly. By comparison, a tree's surface area is much smaller than its volume, so it needs more time to heat up before it ignites.

As the fire progresses, it dries out the material just beyond it——heat and smoke approaching potential fuel causes the fuel's moisture to evaporate. This makes the fuel easier to ignite when the fire finally reaches it. Fuels that are somewhat spaced out will also dry out faster than fuels that are packed tightly together, because more oxygen is available to the thinned-out fuel. More tightly-packed fuels also retain more moisture, which absorbs the fire's heat.

Wind and Rain

Weather plays a major role in the birth, growth and death of a wildfire. Drought leads to extremely favorable conditions for wildfires, and winds aid a wildfire's progress —— weather can spur the fire to move faster and engulf more land. It can also make the job of fighting the fire even more difficult. There are three weather ingredients that can affect wildfires:

Temperature

Wind

Moisture

As mentioned before, temperature has a direct effect on the sparking of wildfires, because heat is one of the three pillars of the fire triangle. The sticks, trees and underbrush on the ground receive radiant heat from the sun, which heats and dries potential fuels. Warmer temperature allow for fuels to ignite and burn faster, adding to the rate at which a wildfire spreads. For this reason, wildfires tend to rage in the afternoon, when

9. In a wildfire, the main variable in ignition time is the ration of the fuel's total surface area _____. （to its volume）

少量的燃料，也叫做引火燃料，如干草、松针、干叶子、嫩枝和其他的死树比大型的原木、树桩燃得快（这就是为什么你点火使用引火物而不是原木）。从化学层面上讲，不同的燃料比其他物质更难点燃。但在森林大火中，大多数燃料是由一种物质组成的着火时间的变化主要是燃料总体的表面积和体积的比。因为一根嫩叶的表面积比它的体积小一点，所有它点燃得快。相比之下，树木的表面积比它的体积小得多，所以点燃它要花更多时间。

随着火势发展，火会蒸干在其上方的燃料——靠近潜在的燃料的热量和烟会引起燃料水气的蒸发，这会使火到达燃料时更容易点燃。燃料之间稍微隔开也会比燃料紧紧绑在一起干得快，因为这些有间隔的燃料更易获得氧气，贴得更紧的燃料也更容易保留吸收热量的水气。

风和雨

天气对森林大火的生成、蔓延和扑灭起主要的作用。干旱是最易引发火灾的条件，风有助于火的发展——天气能加速火势燃烧、扩大面积，也给救火工作带来更大的困难。能够影响森林大火的气候因素有：

温度

风力

湿度

如上所述，温度对火灾的引发有直接的关系，因为热量是火三角形的其中一个因素。树枝、树木及丛林吸取太阳的辐射热，对潜在的可燃物加热和干燥，严热的温度使可燃物被引燃并快速燃烧，加快火势发展的速度。介于这样的原因，火灾多在下午发生，此时的气温是最高的。

风力对火势的影响可能是最大的。它也是一个不确定的因素。风为大火提供充足的氧气，进一步干化潜在的可燃物以及推动火焰以更快的速度蔓延过地面。

temperatures are at their hottest.

Wind probably has the biggest impact on a wildfire's behavior. It's also the most unpredictable factor. Winds supply the fire with additional oxygen, further dry potential fuel and push the fire across the land at a faster rate.

10. As far as a fire's behavior is concerned, the most unpredictable factor is _____. (**wind**)

The stronger the wind blows, the faster the fire spreads. The fire generates wind of its own that are as many as 10 times faster than the wind of surrounding area. It can even throw embers into the air and create additional fires, an occurrence called spotting. Wind can also change the direction of the fire, and gusts can raise the fire into the trees, creating a crown fire.

细节
确认

5. The wind generated by the fire itself are a little slower than the wind of the surrounding area. （**N** ）

While wind can help the fire to spread, moisture works against the fire. Moisture, in the form of humidity and precipitation, can slow the fire down and reduce its intensity. Potential fuels can be hard to ignite if they have high levels of moisture, because the moisture absorbs the fire's heat. When the humidity is low, meaning that there is a low amount of water vapor in the air, wildfires are more likely to start. The higher the humidity, the less likely the fuel is to dry and ignite.

Since moisture can lower the chances of a wildfire igniting, precipitation has a direct impact on fire prevention. When the air becomes saturates with moisture, it releases the moisture in the form of rain. Rain and other precipitation raise the amount of moisture is fuel, which suppresses any potential wildfires from breaking out.

风越大，火势蔓延得越快。火本身产生的风比周围区域产生的风快 10 倍。它甚至能将燃烧的灰烬吹到空中从而引起其他的大火，这种现象叫火点。风可以改变大火的方向，并且能将火引上树，造成树冠火。

当风帮助火蔓延时，湿气能控制火焰。湿气，以潮湿和降水的形式使火焰减弱并且减缓强度。潜在的可燃物在高湿度时很难点燃，因为湿气能够吸收火的热量。湿度降低时，意味着空气中的水气含量低，非常容易引起火灾。湿度越高，可燃物干燥和引燃的可能性越小。

由于湿气能降低火灾发生的可能性，因此降雨对火灾的预防有直接的影响。当空气中的湿气开始饱和时，它会以雨的形式释放。雨水和可燃物中大量湿气形成的其他凝结，抑制任何潜在的可能发生的森林大火。

Fire on the Mountain

The third big influence on wildfire behavior is the lay of the land, or topography. Although it remains virtually unchanged, unlike fuel and weather, topography can either aid or hinder wildfire progression. The most important factor in topography as it relates to wildfire is slope.

Unlike humans, fires usually travel uphill much faster than downhill. The steeper the slope, the faster the fire travels. Fires travel in the direction of the ambient

细节
确认

山上的大火

第三个对森林大火趋势有巨大影响的是地形走势或地势。虽然地势实际上是不变的，不像燃料和天气，但它可以帮助或隐藏火的进行。地势中最重要的与森林大火相关的因素是斜面。

6. Fires are like humans in that they usually travel downhill much faster than uphill. (N)

wind, which usually flows uphill. Additionally, the fire is able to preheat the fuel further up the hill because the smoke and heat are rising in that direction. Conversely, once the fire has reached the top of a hill, it must struggle to come back down because it is not able to preheat the downhill fuel as well as the uphill.

火跟人不一样，它通常上山比下山快。斜面越陡，火爬得越快。火是顺着周围的风的方向蔓延的，通常这些风是上山风。另外，火能预热更高山上的燃料，因为烟和热量是顺着那个方向上升的。相反地，火一旦到达山顶，就必定很快往下滑，因为它不能像预热山上的燃料那样预热山下的燃料。

In addition to the damage that fires cause as they burn, they can also leave behind disastrous problems, the effects of which might not be felt for months after the fire burns out. When fires destroy all the vegetation on a hill or mountain, it can also weaken the organic material in the soil and prevent water from penetrating the soil. One problem that results from this is extremely dangerous erosion that can lead to debris flows.

细节
确认

 7. In fact, the effects of fires might not be felt for months. (Y)

除了火燃烧时引起的损失外，还留下巨大的灾难性问题，它们的影响在燃烧后数月可能不会察觉到。当火毁坏了小山或大山的所有植被时，也损害了土壤中的有机物，并阻止水渗入土壤。这产生的一个问题就是极其危险的侵蚀，可能导致泥石流。

While we often look at wildfires as being destructive, many wildfires are actually beneficial. Some wildfires burn the underbrush of a forest, which can prevent a larger fire that might result if the brush were allowed to accumulate for a long time. Wildfires can also benefit plant growth by reducing disease, releasing nutrients from burned plants into the ground and encouraging new growth.

我们通常看到的是森林大火破坏性的一面，事实上很多森林大火是有益的。大火烧掉森林里的丛林，从而阻止了长时间积累的丛林引起更大的火灾。大火也有益于植物的生长，因为它减少了疾病的传播，从燃烧后的植物中释放营养成分到土壤，促进生长。

【答案解析】

1.【题意】本文讲的是野火如何产生和持续。

　　【解析】首先看文章标题 Wildfires，再浏览各小标题，就可知本文的大概内容。故本题应选的是 YES。叙述符合原文。

2.【题意】木头的燃点应该是 300°F。

　　【解析】本句讲的是木头燃烧所需要的温度。在文中 Fire starts 部分可以查到，Wood's flash point is 572 degrees Fahrenheit (300℃)，所以题目的叙述与原题不符。木头的燃点应该是 572 degrees Fahrenheit。故此题选 NO。

3.【题意】引起火灾的主要原因是人类，不是自然。

　　【解析】本句讲的是引起火灾的主体，因此可以锁定在 Fueling the Fames 小标题下面，该段最后一句话由 However 引起的转折句中，the majority of wildfires are the result of human carelessness. 可知火灾的主要原因是由于人类的疏忽大意，故本题答案是 YES。

4.【题意】在很多情况下，我们以吨来衡量每公顷的可燃物负荷。

　　【解析】本题的关键词是 fuel load，我们把目标锁定在 Wind and Rain 部分，可见这句话：Fuel load is measured by the amount of available fuel per unit area, usually tons per acre. 由此可知，衡量的标准是英亩而非公顷。故本题应选的是 NO。

5.【题意】火本身产生的风比周围区域的风要慢。

　　【解析】有题干可知，答案应在 Wind and Rain 部分，本题讲的是火本身产生的风比周围区域的风要慢。根据文中 The fire generates wind of its own that are as many as 10 times faster than the wind of surrounding area. 可知此题叙述不符合原文，故本题应选的是 NO。

6.【题意】火就像人类一样，通常下山比上山速度要快得多。

【解析】本题提到，火就像人类一样，通常下山比上山速度要快得多。我们可将答案锁定在 Fire on the Mountain 小标题下，该部分第二句：Unlike humans, fires usually travel uphill much faster than downhill. 题目的叙述与原文不符，故本题应选的是 NO。

7.【题意】事实上，火灾的影响或许数月都不会被感觉到。

【解析】本题涉及 the effects of fires 的问题，通过查找可知…the effects of which might not be felt for months after the fire burns out. 与题目所表达的意思相符，故本题应选 YES。

8.【题意】点燃和燃烧的三要素分别是：燃料、空气和热源。

【解析】此题为细节题。题目涉及的是 three components needed for ignition and combustion (点燃和燃烧的三要素)。由此可将目标锁定在此句中：There are three components needed for ignition and combustion to occur. A fire requires fuel to burn, air to supply oxygen, and a heat source to bring the fuel up to ignition temperature. 可见本题要填写的是 a heat source。

9.【题意】在野火中，燃点中的主要变量是燃烧物总燃烧面积达到的整个体积比率。

【解析】由关键词 variable in ignition time (燃点的变量) 可以锁定目标在 Fueling the Flames 小标题，仔细查找可以看到这句话：…the main variable in ignition time is the ratio of the fuel's total surface area to its volume. 所以此处可填 to its volume。

10.【题意】火灾中最难预见的因素是风。

【解析】本题讲的是火灾中最难预见的因素。可以在 Wind and Rain 小标题部分看到：Wind probably has the biggest impact on a wildfire's behavior. It's also the most unpredictable factor. 其中 probably has the biggest impact on a wildfire's behavior 就是对此题目的同义转述，可知本题该填写的答案为：wind。

Part III Listening Comprehension
Section A

11. M: How much are these jackets?

W: They are on sale today, sir. Twenty-five dollars each, or two for forty dollars.

Q: How much does one jacket cost?

【答案及解析】本题属于推理题。男士问："这些夹克衫多少钱？" 女士说："今天削价处理，25 美元一件，40 美元两件。" 可推出正确答案是 A）Twenty-five dollars. 25 美元一件。

12. M: Shall we have dinner in that French restaurant?

W: I can't eat a thing. I feel too bad. My stomach aches.

Q: What do you think the woman will do?

【答案及解析】 本题属于推理题。男士问："今天我们吃法式西餐吧？"女士回答："我一点都吃不下，我感觉非常不舒服，我胃疼。" 可推出正确答案是 D) To stay at home. 呆在家。

13. M: Tina's husband is friendly and easy-going.

W: Yes, just the exact opposite to her brother.

Q: What is Tina's brother like?

【答案及解析】 本题属于推理题。男士说："Tina 的丈夫既友好又随和。"女士说："是的，和他的弟弟正相反。" 可推出正确答案是 D) Not easy-going. 不随和。错误选项：A) Easy-going and friendly. 友好随和。B) Very nervous.很紧张。C) Angry. 爱生气。

14. W: Last night, we went to Peter's house to listen to music.

M: I heard that he has more than 300 jazz records. Is that right?

Q: What do we learn from the conversation?

【答案及解析】 本题属于推理题。女士说："昨晚，我们去彼得家听音乐了。"男士说："我听说他有 300 多张爵士乐唱片，是真的吗？" 可推出正确答案是 B) He is a jazz fan.他是爵士乐迷。A) He plays jazz music. 他演奏爵士乐。属于干扰项；C) He needs 300 jazz records. 他需要 300 张爵士乐唱片。D) He likes classical music. 他喜欢古典乐。属于错误选项。

15. M: I need six stamps and I'd like to send these two books by air mail.

W: Here are your stamps, but you have to go to the next window for the books.

Q: Where does this conversation most probably take place?

【答案及解析】 本题属于推理题. 男士说："我需要六张邮票，我想空邮这两本书。"女士说："这是邮票，但是你得到下一窗口邮书。" 可推出正确答案是 A) At a post office. 在邮局。

16. W: Jim, would you mind driving me to my school?

M: Sure, why not?

Q: How does Jim respond to the woman?

【答案及解析】 本题属于推理题。女士说："吉姆，你愿意开车送我去学校吗？"男士说："当然，为什么不？"可推出正确答案是 D) He is glad to drive her there. 他很乐意送她去那儿。错误选项：A) He was scared. 他害怕了。B) He was upset. 他很不开心。C) He hasn't got a car. 他没车。

17. M: Did you see Mary in the business office?

W: Yes, she was applying for a student loan.

Q: What was Mary doing?

【答案及解析】 本题属于推理题。男士说："你在办公室看见玛丽了吗？"女士说："是的，她在申请助学贷款。"可推出正确答案是 D) Asking for some financial aid. 请求经济上的援助。错误选项：A) Lending money to a student. 借钱给一个学生。B) Filling a form. 正在填写一张表格。C) Reading a student's application. 读一个学生的申请表。

18. W: When is the movie to start?

M: Don't worry. It doesn't start until 12:30, we've still got 30 minutes.

Q: What's the time now?

【答案及解析】 本题属于推理题。女士说："电影什么时候开始？"男士说："别担心，电影 12:30 开始，我们还有 30 分钟。" 可推出正确答案是 C) 12:00。

Now you will hear two long conversations.

Conversation One

W: Hello, Gary. How are you?

M: Fine! And yourself?

W: Can't complain. Did you have time to look at your proposal?

M: No, not really. Can we go over it now?

W: Sure. I've been trying to come up with some new production and advertising strategies. First of all, if we want to stay competitive, we need to modernize our factory. New equipment should've been installed long ago.

M: How much will that cost?

9

W: We have several options ranging from one hundred thousand dollars all the way up to half a million.

M: OK. We'll have to discuss these costs with finance.

W: We should also consider human resources. I've been talking to personal as well as our staff at the factory.

M: And what's the picture?

W: We'll probably have to hire a couple of engineers to help us modernize the factory.

M: What about advertising?

W: Marketing has some interesting ideas for television commercials.

M: TV? Isn't that a bit too expensive for us? What's wrong with advertising in the papers, as usual?

W: Quite frankly, it's just not enough any more. We need to be more aggressive in order to keep ahead of our competitions.

M: Will we be able to afford all this?

W: I'll look into it, but I think higher costs will be justified. These investments will result in higher profits for our company.

M: We'll have to look at the figures more closely. Have finance draw up a budget for these investments.

W: All right. I'll see to it.

Questions 19 to 22 are based on the conversation you have just heard.

19. What are the two speakers talking about?

【解析】细节题。从 I've been trying to come up with some new production and advertising strategies. 一句可知女士的报告主要是关于新产品和广告推介的策略问题，即新商业策略的建议。可知答案 D)正确。

20. What does the woman say about the equipment of their factory?

【解析】推理判断题。从 New equipment should've been installed long ago. 一句可知说话者所持的态度。可知答案 B)正确。

21. What does the woman suggest about human resources?

【解析】细节题。从 We'll probably have to hire a couple of engineers to help us modernize the factory. 可知女士建议聘请一些工程师来提高工厂的现代化水平。可知答案 B)正确。

22. Why does the woman suggest advertising on TV?

【解析】态度明示题。从 "Quite frankly"可知后面一句为说话者个人的观点，她认为报纸广告 "not enough"。应添加，可知答案 C)正确。

Conversation Two

W: Sir, you've been using the online catalogue for quite a while. Is there anything I can do to help you?

M: Well, I've got to write a paper about Hollywood in the 30s and 40s, and I'm really struggling. There are hundreds of books, and I just don't know where to begin.

W: Your topic sounds pretty big. Why don't you narrow it down to something like…uh…the history of the studios during that time?

M: You know, I was thinking about doing that, but more than 30 books came up when I typed in "movie studio".

W: You could cut that down even further by adding the specific years you want. Try adding "1930s" or "1940s"or maybe "Golden Age."

M: "Golden Age" is a good idea. Let me type that in…Hey, look, just 6 books this time. That's a lot better.

W: Oh…another thing you might consider…have you tried looking for any magazine or newspaper articles?

M: No, I've only been searching for books.

W: Well, you can look up magazine articles in the Reader's Guide to Periodical Literature. And we do have the Los Angeles Times available over there. You might go through their indexes to see if there's anything you want.

M: Okay. I think I'll get started with these books and then I'll go over the magazines.

W: If you need any help, I'll get started with these books and then I'll go over the magazines.

W: If you need any help, I'll be over at the Reference Desk.

M: Great, thanks a lot.

Questions 23 to 25 are based on the conversation you have just heard.

23．What is the man doing?

【解析】细节归纳题。从 I've got to write a paper about Hollywood in the 30s and 40s, and I'm really struggling. 及后面出现的词语如 Reference Desk 可归纳出男士是在图书馆找资料。可知答案 A)正确。

24．What does the libertarian think of the topics the man is working on?

【解析】态度判断题。从 Why don't you narrow…一句可知图书馆员认为男士的论文题材太宽泛了。可知答案 A)正确。

25．Where can the man find the relevant magazine articles?

【解析】细节题。从 you can look up magazine articles in the Reader's Guide to Periodical Literature.一句可知答案 D)正确。

Section B

Passage One

The Yale University professors agreed in a panel discussion tonight that the automobile was what one of them called "Public Health Enemy No.1 in This Country". Besides polluting the air and congesting the cities, automobiles could cause heart disease "because we don't walk anywhere any more", said Dr. H. P. Richard Weinerman, professor of medicine and public health. Dr. Weinerman's sharp indictment of the automobile came in a discussion of human environment on Yale Reports, a radio program broadcast by Station WTIC in Hartford, Connecticut. The program opened a three part series on "Staying Alive". "For the first time in human history, the problem of man's survival has to do with his control of man-made hazards," Dr. Weinerman said. Before this, the problem had been the control of natural hazards.

Questions 26 to 28 are based on the passage you've just heard.

【内容概要】本文讲述的是汽车对人类健康的威胁。耶鲁大学的教授们在会议上一致认为汽车是威胁大众健康的头号敌人。汽车污染空气，挤塞城市，而且，汽车使人们以车代步，从而引发疾病。人类史上第一次把人类生存问题与控制人类自产公害相联系起来，而在此前提及的是控制自然灾害。

26．What is the main idea of the passage?

【解析】主旨题。文章第一句即为主题句，此句的后半句 "…the automobile was what one of them called "Public Health Enemy No.1 in This Country" 点明了文章中心：汽车是公众健康的头号敌人。因此选项 C （汽车威胁着人的健康）是正确答案。

27．Why could automobiles cause heart disease?

【解析】细节题。从文章中的第二句话，"…because we don't walk anywhere any more，…" 可知以车代步容易引发疾病，影响人的身体健康。因此选项 D 是正确答案。

28．For the 1st time, what does the problem of man's survival have to do with?

【解析】细节题。只要听到 "For the first time in human history, the problem of man's survival has to do with his control of man-made hazards." 一句就知道人类是第一次处理人造公害的控制问题。因此选项 D 是正确答案。

Passage Two

A small dog brought disaster to a small English town last week. It was trying to cross a busy street but was too frightened to leave the strip in the middle of the street. A truck driver parked his truck on the side of the road and got out to help it. While he was going to get the dog, his truck rolled down the street. It smashed into four parked cars, crashed through a fence, rolled down a bank and smashed into a row of houses. Only the driver was hurt. The dog bit him on the hand while he was carrying it.

Here is another story about dogs. A nine-year-old child who nearly drowned while she was swimming in a home swimming pool, was saved by the family dog on Saturday. The child was alone in the large pool at the time.

The family of the dog said it would receive an extra large bone as a reward.

Questions 29 to 32 are based on the passage you have just heard.

【内容概要】本文讲述的是两只狗的故事。上周，有一只小狗在路中央不肯挪动，一位司机因帮它，引出了一系列事故。幸运的是，只是司机受伤了，而且还是在抱狗时被狗咬的。而另外一只狗救了一个在家庭游泳池溺水的九岁小孩。

29．What happened to the dog that brought disaster to the town?

【解析】细节题。从文章的第二句话，"…but was too frightened to leave the strip in the middle of the street." 可知，狗因过害怕而在路中央不敢移动。因而选项 C 为正确答案。

30．Who was hurt in the disaster?

【解析】细节题。从文章中 "Only the driver was hurt," 一句可知是卡车司机受伤了。选项 C 即为正确答案。

31．Where did the second story about dog take place?

【解析】细节题。从文章中此句 "…while she was swimming in a home swimming pool, was saved…"，可知故事发生在家庭游泳池。选项 B 为正确答案。

32．What would be the reward given to the dog in the second story?

【解析】细节题。从文章中最后一句 "…said it would receive an extra large bone as a reward." 可知这条狗所得的奖励是主人要额外地给它一大块骨头。选项 D 为正确答案。

Passage Three

If you are like most people, your intelligence varies from season to season. You are probably a lot sharper in the spring than you are at any other time of year.

A noted scientist, Ellsworth Huntington, concluded from other men's work and his own among people in different climate and temperature have a definite effect on our mental abilities.

He found that cool weather is much more favorable for creative thinking than is summer heat. This does not mean that all people are less intelligent in summer than they are during the rest of the year. It does mean, however, that the mental abilities of large numbers of people tend to be lowest in summer.

Spring appears to be the best period of the year for thinking. One reason may be that in the spring man's mental abilities are affected by the same factors that bring about great changes in all nature.

Fall is the next best season, then winter. As for summer, it seems to be a good time to take a long vacation from thinking.

Questions 33 to 35 are based on the passage you've just heard.

【内容概要】本文讲述的是多数人的智力发挥会受季节的影响。季节不一样，人的智力发挥也会不一样，大多数人都会这样。春天是创造性思维最活跃的季节，其次是秋天，再而是冬天。而夏天的酷热会使人的思维发展最慢。当然，并不是所有人都这样。

33. What is the passage mainly about?

【解析】主旨题。文章中的第一句就点明了主题，"…your intelligence varies from season to season." 可知答案 B 正确。

34. What is the best season for thinking?

【解析】细节题。从文章中句子 "Spring appears to be the best period of the year for thinking." 可知答案 D 正确。

35. Which of the following statements is true according to the passage?

【解析】综合判断题。从文章中句子 "This does not mean that all people are less intelligent in summer…" 可知选项 A 不对。文章中明确提及 "summer heat" 对人的智力发挥有影响，故选项 B 也不对。选项 C 没有提及。而从此句 "A noted scientist concluded…that climate and temperature have a definite effect on mental ability." 可知选项 D 正确。

Section C

Very high waves are (36) destructive when they strike the land. Fortunately, this (37) seldom happens. One reason is that out at sea, waves moving in one direction almost always run into waves moving in a different direction. The two (38) sets of waves tend to cancel each other out. Another reason is that water is (39) shallower near the shore. As a wave gets closer to land, the shallow (40) bottom helps reduce its (41) strength.

But the power of waves striking the (42) shore can still be very great. During a winter gale, waves sometimes (43) strike the shore with the force of 6,000 pounds for each square foot. That means a wave, 25 feet high and 500 feet along its face, may (44) strike the shore with a force of 75 million pounds.

Yet (45) the waves, no matter how big or how violent, affect only the surface of the sea. During the most raging storms, (46) the water a hundred fathoms (600 feet) beneath the surface is just as calm as on the day without a breath of wind.

Part IV Reading Comprehension (Reading in Depth)
Section A

【短文大意】第一届现代奥林匹克运动会于 1896 年在雅典举行，当时只有 12 个国家参加。除了东道国之外，许多参赛者都是当时碰巧在希腊观光的旅游者。尽管整个比赛是非正式的，而且标准也不高，但奥运会作为业余体育运动的古老传统却被保持下来了。

从那时起，奥运会每四年举办一次，不过在两次世界大战期间，奥运会停办了。毫无疑问，这与古老的奥林匹克精神是背道而驰的，因为在那时，战争必须停下来为比赛让路。

奥运会发展规模巨大，体育成绩已经达到了前所未有的高度。不幸的是，奥运会的道德水准没有得到相应提高。现代奥运会已经不在奥林匹亚而是在全世界各个不同城市举行。由于举办奥运会带来的政治威望和商业利润，各国为争夺举办奥运会而相互竞争，这样就使政治和商业利益不可避免地卷入进来。第 11 届奥运会于 1936 年在柏林举行，刚在德国当权的希特勒试图利用这个大好时机进行纳粹宣传。于是奥林匹克圣火第一次以接力方式——现在需要好几个月才能完成的一段马拉松路程——从奥林匹亚一直传送到举办地点。

47.【答案及解析】I) 空格中要填的词为 who 引导的定语从句中的谓语动词，可知此处应填动词的过去式。可选项有 had, occurred 和 happened，空格所在的 who 引导的定语从句修饰的是 tourists "游客"，happened to be in Greece "碰巧在希腊" 是符合这样的语境的，而 had to "不得不" 和 occur to 表示 "想起，想到" 都不能使句意通顺。

48．【答案及解析】L) and 引导两个意思相近的并列结构，由 the standard was not high "标准不高" 可知，此处要填的是和 not high 语气一致的形容词，再由前两句可知，参赛的国家很少并且有很多游客参加了比赛，这说明这次比赛没有那么正式，选项中只有 informal 符合题意。选项 irregular 则一般表示 "(形状)不规则的，(安排)无规律的"，与句意不符，故排除。

49．【答案及解析】M) 从句中的 except 可知在两次世界大战中没有比赛，说明奥运会因战争而中断。选项中有 end 和 interruption 可表示停止，end "终断"，表示某事中断后不再继续，interruption "中断"，表示某事中断后仍然继续，故选择 interruption。

50．【答案及解析】J) 本句主系表结构完整，由此可知，此处应填副词，修饰整个句子。选项中有 definitely 和 especially，从句意来理解，这……违背了古老的奥林匹克精神，将 definitely "显然地" 带入原文，符合原文的意思；而 especially "特别地" 表示强调，而此处表示的则是通过比较后得出结果，故排除 especially。

51．【答案及解析】K) 奥林匹克比赛中两个最重要的方面就是竞技水平和运动精神，由 grown enormously in scale 可知，现在参加奥林匹克运动会规模渐大，运动员逐渐增多，因此应理解为体育竞技水平达到了前所未有的高度，因此应选 physical。

52．【答案及解析】D) 此处应填形容词。由 Unfortunately 和 because of the political prestige and commercial profit 可推出，不幸的是各国竞争举办奥运会的目的是为了政治威望与利益，可见此处说明的是与体育成绩相对应的运动会的道德水准没有得到相应提高，选项中只有 moral 符合题意。

53．【答案及解析】A) 此处应填名词。各国争夺举办奥运会，to hold the games 是一种实力象征，也是一种荣耀。选项中有 honor 和 power 可选，power 一般指 "(个人或团体所掌握或获得的)职权"，故排除 power 而选 honor "荣幸"。

54．【答案及解析】C) 此处应填动词的被动形式，这里考查的是动词 have 的一个搭配用法，即 have something out of something "从……中得到……"，故这里选择 C)had，表示 "举办奥运会带来的政治威望和商业利润。"

55．【答案及解析】H) come to power 是固定搭配，意为 "(开始)掌权，上台"。

56．【答案及解析】B) 此处应填动词的原形。可选项有 accomplish, arrive 和 end，arrive 一般指 "到达(某地)"，end 强调的是 "(使某事)结束"，因此都不能与 journey 搭配，故排除 arrive 和 end 而选 accomplish "完成"。

Section B

Passage one

A subject which seems to have been insufficiently studied by doctors and psychologists is the influence of geography and climate on the psychological health of humankind. There seems no doubt that the general character of landscape, the relative length of day and night, and climate must all play a part in determining what kind of people we are.

It is true that a few studies have been made. Where all the inhabitants of a particular area enjoy exceptionally good or bad health, scientists have identified contributory factors such as the presence or absence of substances like iodine, fluoride, calcium, or iron in the water supply, or

57. The author's purpose of writing this passage is to _____.

主旨
大意

A) alert readers to the scarcity of natural resources

B) call for more research on the influence of geographical environment

C) introduce different elements in character cultivation

D) draw more attention to the health condition of mankind

58. It can be inferred that proper amounts of iodine, fluoride and calcium can _____.

细节
推断

A) benefit people's physical health

B) influence the quality of water supply

C) help provide breeding places for pests

D) strengthen a person's character

59. How does the author evaluate the generalizations of people's types in Para. 3?

perhaps types of land that provide breeding places for pests like mosquitoes or rats.

Moreover, we can all generalize about types of people we have met. Those living in countries with long dark winters are apt to be less talkative and less lively than inhabitants of countries where the climate is more equable. And where olives and oranges grow, the inhabitants are cheerful, talkative, and casual.

But these commonplace generalizations are inadequate—the influence of climate and geography should be studied in depth. Do all mountain dwellers live to a ripe old age? Does the drinking of wine, rather than beer, result in a sunny and open character? Is the strength and height of a Kenyan tribe due to their habitual drinking of the cow blood?

We are not yet sure of the answer to such questions, but let us hope that something beneficial to humankind may eventually result from such studies.

细节推断

A) Such generalizations help us judge the different characters of people we meet.

B) Such generalizations are not inclusive enough to draw a convincing conclusion.

C) Such generalizations prove that nature plays an important role in determining social habits.

D) Such generalizations show that there are mainly two different types of people on the planet.

60. According to the passage, research into the influence of climate and geography should _____.

A) focus on unknown aspects

B) be pursued on a larger scale

C) be carried out among remote tribes

细节辨认 D) go ahead in depth

61. What do we know about the generalizations of people's type?

A) People who like drinking wine tend to be optimistic.

B) People who live in mountain areas tend to have a long life.

C) People who live in areas with stable climate tend to be talkative and lively.

D) People who like drinking cow blood tend to be strong and tall.

【短文大意】 对于地理和气候对人类心理和身体健康的影响这个问题，医生和心理学家还研究得很不充分。毫无疑问，地形的综合特性、昼夜的相对长短和气候等因素都对决定我们是什么类型的人起作用。

有关的研究的确做了一些。个别地区居民的身体异常地好或坏，科学家已经确认这和供水中有无碘、氟、钙或铁等元素有关，也可能和土地类型适于有害动物例如蚊子和鼠类的滋生有关。

此外我们都能概括我们所遇到的人的大概类型。居住在有漫长黑暗冬季的国度的居民，比那些气候较稳定的国家居民少言寡语，不太活跃，橄榄树和橘子生长的地方，居民则快乐、健谈和漫不经心。

但是这种一般性的归纳并不充分，应该更深入地研究气候和地理对人的影响。所有的山地居民都能活到高龄吗？喝葡萄酒而不是啤酒能产生开朗和外向的性格吗？肯尼亚部落人强壮是由于他们习惯喝奶牛血吗？

我们还不能确定这类问题的答案，但是我们希望这类研究最终能产生有利于人类的结果。

57.【题目译文】 作者写这篇文章的目的是_____。

【答案及解析】B) 本文一开头就提出：关于地理环境对人类的影响，研究还不充分。接下来的每段开头句也都进一步阐述这个观点，如第2段首句和第4段首句，结尾句又提出希望，所以正确答案是B)。

58.【题目译文】 可以推断：适量的碘、氟和钙能够_____。

【答案及解析】A) 见文中第2段第2句"Where all the inhabitants of a particular area enjoy exceptionally good or bad health, …"句中的good health 与后面的contributory factors 告诉读者选项A)是正确的。

59.【题目译文】 作者在第3段是如何评价对人的类型的概括的？

60.【题目译文】 根据本文，对气候和地理影响的研究应该_____。

【答案及解析】D) 见文中第 4 段第 1 句破折号后 "the influence of climate and geography should be studied in depth" 指出应对地理和气候的影响进行更深入的研究，所以选项 D)是正确的。

61.【题目译文】 我们知道哪些关于人的类型的概括？

【答案及解析】C) 见文中第 3 段第 2 句 "Those living in countries with long dark winters…"，其中 equable 与选项C)中的 stable 是同义词转换，指出那些气候较稳定的国家居民比较健谈、活跃，所以选项 C)是正确的。

Passage Two

To forgive may be divine, but no one said it was easy. When someone has deeply hurt you, it can be extremely difficult to let go of your complaints and hatred. But forgiveness is possible — and it can be surprisingly beneficial to your physical and mental health.

"People who forgive show less depression, anger and stress and more hopefulness," says Frederic, Ph.D., author of *Forgive for Good*. "So it can help save on the wear and tear on our organs, reduce the wearing out of the immune system and allow people to feel more vital."

So how do you start the healing? Try following these steps:

Calm yourself. To remove your anger, try a simple stress-management technique, "Take a couple of breaths and think of something that gives you pleasure: a beautiful scene in nature, someone you love," Frederic says.

Don't wait for an apology. "Many times the person who hurts you has no intention of apologizing," Frederic says. "They may have wanted to hurt you or they just don't see things the same way. So if you wait for people to apologize, you could be waiting an awfully long time." Keep in mind that forgiveness does not necessarily mean reconciliation with the person who upset you or neglecting his or her action.

Take the control away from your offender. Mentally replaying your hurt gives power to the person who caused you pain. "Instead of focusing on your wounded feelings, learn to look for the love, beauty and kindness around you," Frederic says.

Try to see things from the other person's perspective. If you empathize with that person, you may realize that

62. By saying that forgiveness "can help save on the wear and tear on our organs," Frederic, Ph. D. means that _____.

A) people are likely worn out by crying when they get hurt

B) we may get physically damaged if we stick to the hurt

C) our physical conditions benefit most from forgiveness

D) the immune system is closely related with our organs

63. When you try to calm yourself, you are actually trying to_____.

A) recall things you love

B) show you are angry

C) relieve your stress

D) breathe normally

64. Your offender may not want to apologize because _____.

A) they are afraid that they won't be forgiven

B) they don't even realize they have hurt you

C) they don't share the same feeling with you

D) they think that time can heal any wound

65. You will still be under the control of the offender if _____.

A) the offender refuses to reconcile with you

B) you keep reminding yourself on the pain

he or she was acting out of ignorance, fear — even love. To gain perspective, you may want to write a letter to yourself from your offender's point of view.

Recognize the benefits of forgiveness. Research has shown that people who forgive report more energy, better appetite and better sleep patterns.

Don't forget to forgive yourself. "For some people, forgiving themselves is the biggest challenge," Frederic says. "But it can rob you of your self-confidence if you don't do it."

C) the offender never feels sorry to you

D) you don't find love, beauty or kindness

细节
推断

66. Which of the following enables you to gain the offender's perspective?

A) Empathizing with the offender.

B) Realizing the reason for the offender's action.

C) Writing a letter to the offender.

D) Doing the same thing the offender did to you.

【短文大意】宽恕可能很神圣，但没有人说过这是容易做到的。如果有人深深地伤害了你，要你不抱怨、不憎恨可能极其不易。但是宽恕是可以做到的——而且会给你的身心健康带来意想不到的好处。

"Forgive for Good" 一书的作者弗雷德里克博士说："宽恕的人表现得不那么压抑、愤怒、紧张，并且怀有更多的希望。因此宽恕可以有助于减少器官的损耗，降低免疫系统的消耗，使人们感觉更有活力。"

那么如何开始疗伤呢？试试下面的步骤：

让自己平静下来。尝试一个简单的减压技巧来消除心中的怒火。弗雷德里克说："呼吸几次，想想给你带来乐趣的东西：如大自然的美景、你爱的人。"

不要等待别人道歉。弗雷德里克说："很多时候，伤害你的人无心道歉，可能是他们有意伤害你，或者只是他们看待问题的方式与你不同而已。因此，如果你等待别人来道歉，你可能等很长一段时间。"记住：宽恕不一定是向伤害你的人妥协，或是原谅他/她的行为。

不要让伤害你的人控制你的情绪。心里反复想象所受到的伤害，会让伤害你的人伤你更深。"不要陷入受伤的情感而不能自拔，学会寻找身边的爱、美、善。"费雷德里克说。

试着从对方的角度看问题。如果你站在对方的立场，你可能会发现，他/她的行为是出于无心、害怕——甚至是爱。要得到这样的认识，你可能要从伤害你的人的角度给自己写封信。

认识到宽恕的好处。有研究显示，宽恕的人精力更充沛，胃口更好，睡觉更香。

不要忘记宽恕自己。费雷德里克说："对有些人来说，宽恕自己是最大的挑战。但是如果你不这样做，你就会丧失自信。"

62.【题目译文】弗雷德里克医生说宽恕"可以有助于减少器官的损耗"，他的意思是＿＿＿＿。

【答案及解析】B) 见文中第 2 段末句 "So it can help save on the wear and tear on our organs, ..." 其中 organs 和 immune system 都表明该句说的是宽恕和身体状况之间的关系，选项 B)将原文内容从反面来说，是本题答案。

63.【题目译文】当你努力想让自己平静下来时，你实际上是在努力去＿＿＿＿。

【答案及解析】C) 见原文第 4 段第 2 句的暗示 "To remove your anger, try a simple stess-management technique, ..."，说明消除怒火使自己平静下来属于 stress management 的范畴，可见选项 C)是正确的。

64.【题目译文】伤害你的人可能不想道歉，因为＿＿＿＿。

【答案及解析】C) 见原文第 5 段第 3 句"They may have wanted to hurt you or they just don't see things the same way."该句列出了伤害你的人不会道歉的两个原因，选项 C)是第二个原因的近义表达。

65.【题目译文】伤害你的人将仍旧控制你的情绪如果＿＿＿＿。

【答案及解析】B) 选项 B)中的 keep remind yourself on...是对第 6 段第 2 句 mentally replaying...的同义改写，因此选项 B)是本题答案。

66.【题目译文】下列哪一项能够让你从伤害你的人的角度看问题？

【答案及解析】A) 见原文倒数第 3 段首句 "Try to see things from the other person's perspective." 和第 2 句 If 条件句对第一句的解释，选项 A)意思相近，指出移情，即站在其他人的角度看问题有助于宽恕对方。所以选项 A)是本题答案。

Part V   Cloze

【短文大意】大多数人根本就不了解辛苦的工作，而是担心如何去收集那些他们花钱去动物园看的令人着迷的鸟兽。别人经常问我的一个问题就是：当初我是如何成为动物收藏家的。答案是我一直对动物以及动物园感兴趣。据我父母回忆，我首先会说的不是常规的"妈妈"、爸爸，而是"动物园"这个单词，我一遍一遍地尖叫"动物园"这个词，直到有人为了阻止我而带我去动物园。长大一点的时候，我住在希腊，有了好多的宠物，从猫头鹰到海马；我所有的业余时间都去乡下搜集新的标本，加到我的收集品中。后来我在"城市动物园"呆了一年，作为一名助手，我开始积累很难在家里养的大动物的经验，例如：狮子、熊、鸵鸟。当我离开时，已经积攒了我第一次旅游的钱了。从那以后我经常去旅游。虽然收集的工作不容易而且有很多不满意，但是却是一个吸引热爱动物和旅游的人的工作。

67.【答案】A 逻辑搭配。根据下一句及随后的内容，作者讲的是怎样成为 animal collector 的，因此选 A。

68.【答案】C 固定短语搭配。in the first place 是固定短语，意思是"首先"。此句意思是：别人经常问到的问题之一是当初我是如何成为 animal collector 的。

69.【答案】A 上下文词义搭配。这句话的意思是：作者在牙牙学语时，最早发清楚的音是"zoo"，而不是"妈妈"、"爸爸"，因此应选 clarity "清晰。填入其他选项 emotion(感情)，sentiment(多愁善感)，affection(友爱)都不合逻辑。

70.【答案】B 上下文逻辑。not...but 结构，意思为"不是……而是"。

71.【答案】D 上下文词义搭配。根据后面的 over and over again，应选 repeat。

72.【答案】C 近义词辨析。小孩想去动物园，便不停地发出尖叫声，故选 voice。a shrill voice 与 scream 的意思接近。volume（音量）、noise（噪音）、pitch（音调）均不符合要求。

73.【答案】B 固定词组辨析。shut sb. up 是指让某人住口，为了让孩子停止尖叫，只好带他去动物园。

74.【答案】C 固定词组辨析。a great many 后直接跟可数名词的复数形式；a great/large amount of 后跟不可数名词；只有 a great/large number of 后可以跟可数名词的复数形式。例如：I read a great many English books；A large amount of money is spent on tobacco every year；A great number of civilians were murdered in cold blood.

75.【答案】B 近词义辨析。range from...to ...的意思是"范围从……到……"；vary from...to...是指"从……到……不等"；change/alter from... to... 意思是"从……变成……"。

76.【答案】D 动词辨义。Living 后必须接介词 in，意为"居住"；cultivating 耕种；reclaiming 开垦；只有 exploring 有"探索"的意思。

77.【答案】C 动词词组辨析。add to 相当于 increase，意为"增加"。其余选项后面都不接 to。

78.【答案】A 固定词组辨析。later on 为固定短语，意为"后来"。

79.【答案】D 上下文词义辨析。attendant 服务员；keeper 可理解为"饲养员"，但是 a student keeper 容易被误解为"收留学生的人"；assistant 有"助手"之意。作者一边上学，一边在动物园打工，只能当助手。

80.【答案】D 定语从句用法。which 在此引导定语从句，修饰前面列出的动物。

81.【答案】D 上下文词义。因为钱是在动物园打工挣的，选 successfully 更能体现其含义。

82.【答案】D 动词辨义。finance my first trip 意为"支付我的首次旅行费用"；pay 后接介词 for；其他选项的意思相差甚远。

83.【答案】B 副词辨析。此句为现在完成进行时态，选 regularly 比较贴切。

84.【答案】D 上下文词义。此句是由 though 引导的让步状语从句，应与主句意思相对立。选项中 sorrows 和 disappointments 与主句的 appeal to 相对立，但 sorrows 的分量太重。

85.【答案】C 动词词组搭配。appeal to 为短语，意思是"吸引"。

86.【答案】B 近义词搭配。excursion 短途旅行；journey（从一地到另一地的）长距离旅行，是具体的旅行，指一个过程；travel 泛指所有的旅行、旅游、海外旅行、远足。

Part VI   Translation

87. He continued speaking, _____ (不顾及)my feelings on the matter.

【题目译文】他不顾及我在此事上的感情，继续往下说。

【答案及解析】regardless of 不顾。

88. _____ (到目前为止)everything is all right.

【题目译文】到目前为止，一切都好。

【答案及解析】So far/ Up to now/Until now 到目前为止。

89. Will the train arrive _____(准时)?

【题目译文】火车会准时到达吗？

【答案及解析】on time 准时，注意另外一个词组 in time，其含义是及时。

90. I can't go, she'll go _____me(由她代替我去).

【题目译文】我不能去，由她代替我去。

【答案及解析】instead of 代替，而不是。

91. She seems to do these things _____(有意地).

【题目译文】她似乎是有意地做这些事。

【答案及解析】on purpose 故意地。

大学英语4级考试新题型模拟试卷二答案及详解

参考答案

Part I   Writing

The Importance of Reading Classics

It is widely acknowledged that reading the classics is both important and beneficial to the character development and personal growth of the young people. To me, nothing can bring more joy and happiness than reading those masterpieces created by great figures like Confucius and Cao Xueqin. I believe works like *The Dream in the Red Chamber* and *The Legend of Three Kingdoms* can drastically elevate one's aesthetic taste and deepen the understanding of the glorious history of Chinese culture.

However, the modern society is full of temptations. Compared with TV soap operas, sport events, and video games, classical literary works are old-fashioned and time consuming. In bookstores, "Fast-food" reading materials are replacing classics, and young writers with sensational and "cool" remarks win the support of a large number of fans.

As the salt of this world, we college students should be fully aware of the important role the classics play in broadening one's vision. Therefore, we should start reading and studying the treasuries our ancestors left and absorbing the essence of those classical works. We should also advocate to the public the importance of classics so that an increasing number of general people can enjoy the pleasure of reading.

Part II   Reading Comprehension (Skimming and Scanning)

1. Y   2. N   3. Y   4. NG   5. Y   6. Y   7. N

8. official

9. from the very brink of extinction

10. positive attitude and grass roots support

Part III   Listening Comprehension

Section A

11. C   12. D   13. B   14. D   15. B   16. C   17. A   18. C   19. C   20. C

21. C   22. A   23. D   24. D   25. D

Section B

26. D   27. B   28. C   29. B   30. A   31. C   32. D   33. A   34. B   35. D

Section C

36. effort   37. officials   38. negotiate   39. balanced

40. Competition   41. exchange   42. process   43. environmental

44. These "greenhouse gases" trap heat in the atmosphere and, are blamed for changing the world's climate

45. But currently, nations producing, only 44 percent have approved the protocol. Russia produces about 17 percent of the world's green-house gases

46. To join WTO, a country must reach trade agreements with many trading countries that are also WTO members

Part IV   Reading Comprehension (Reading in Depth)

Section A

47. K: projects   48. M: role   49. A: acting   50. J: offers   51. D: cooperative

52. G: forward   53. F: especially   54. I: information   55. O: victims   56. E: entire

Section B

57. C   58. A   59. C   60. B   61. B   62. D   63. B   64. A   65. C   66. B

Part V   Error Correction

67. hangs →hung   68. from→of   69. that→which

70. lamps^→ were   71. sent→send   72. it→itself

73. as→like   74. revolved→revolving

75. however→whatever   76. being→/

Part VI   Translation

77. Thanks to your help

78. Who on earth told you that?

79. In case of

80. by chance

81. he carried six boxes at a time

答案解析及录音原文

Part I   Writing

The Importance of Reading Classics

It is widely acknowledged that reading the classics is both important and beneficial to the character development and personal growth of the young people. To me, nothing can bring more joy and happiness than reading those masterpieces created by great figures like Confucius and Cao Xueqin. I believe works like *The Dream in the Red Chamber* and *The Legend of Three Kingdoms* can drastically elevate one's aesthetic taste and deepen the understanding of the glorious history of Chinese culture.

However, the modern society is full of temptations. Compared with TV soap operas, sport events, and video games, classical literary works are old-fashioned and time consuming. In bookstores, "Fast-food" reading materials are replacing classics, and young writers with sensational and "cool" remarks win the support of a large number of fans.

As the salt of this world, we college students should be fully aware of the important role the classics play in broadening one's vision. Therefore, we should start reading and studying the treasuries our ancestors left and absorbing the essence of those classical works. We should also advocate to the public the importance of classics so that an increasing number of general people can enjoy the pleasure of reading.

【评论】本文第一段首先阐明作者的观点，即承认阅读名著的诸多好处，并通过具体例子加以证实，然后在第二段列举相反方面，最后一段再次扣题并发出号召。本文条理清晰，正反两面论证，句式结构多变，言简意赅。

【常用句型】

第一段：阐明观点 To me, nothing can bring more joy and happiness than …

第二段：反面论证 However, the modern society is full of temptations. Compared with …

第三段：总结并号召 As the salt of this world, we college students should… Therefore, we should start …

Part II   Reading comprehension (Skimming and Scanning)

【文章及答案解析】本文是介绍拯救少数民族语言的说明文。先说语言消亡并不是很奇怪的事，但现在比过去消亡得快。语言学家在上一世纪就开始了对这一现象的认识。然后分几段介绍了拯救的可能性和需要的费用。接着举了4个具体的事例来说明怎样拯救这些即将消亡的少数民族语言。最后总结了拯救的先决条件，就是要人们采取积极地态度与提供支持。

SAVING LANGUAGE

There is nothing unusual about a single language dying. Communities have come and gone throughout history and with them their language. But what is happening today is extraordinary, judged by the standards of the past. It is language extinction on a large scale. According to the best estimate, there are some 6,000 languages in the world. Of course, about half are going to die out in the course of the next century: that's 3,000 languages in 1,200 months. On average, there is a language dying out somewhere in the world every two weeks or so.

How do we know? In the course of the past two or three decades, linguists all over the world have been gathering comparative data. If they find a language with just a few speakers left, and nobody is bothering to pass the language on to the children, they conclude that language is bound to die out soon. And we have to draw the same conclusion if a language has less than 100 speakers. It is not likely to last very long. A 1999 survey shows that 97 percent of the world's languages are spoken by just four percent of the people.

It is too late to do anything to help many languages, where the speakers are too few or too old, and where the community is too busy just trying to survive to care about their language. But many languages are not in such a serious position. Often, where languages are seriously endangered, there are things that can be done to give new life to them. It is called revitalization.

Once a community realizes that its language is in danger, it can start to introduce measures which can genuinely revitalize. The community itself must want to save its language. The culture of which it is a part must need to have a respect for minority languages. There need to be funding to support courses, materials, and teachers. And there need to be linguists to get on with the basic task of putting the language down on paper. That's the bottom line: getting the language documented—recorded, analyzed, written down. People must be able to read and write down. People must be able to read and write if they and their language are to have a future in

【文章大意】 　　拯救语言

1. The rate at which languages are becoming extinct has increased. （Y）

同义转述

一种语言的消亡并不是很奇怪的事情。通观历史，人类使用自己的语言进行交往。但是用过去的标准评判今天所发生的事情则是很特别的，其中在很大程度上是由于语言的消亡。根据最新的统计，世界上有 6,000 种语言。当然，到下一个世纪会有近半数的语言消失，即 1,200 个月中有 3,000 种语言消失。世界上平均每两周左右就有一种语言消亡。

细节辨认
4. Language extinct more quickly in certain parts of the world than in others. (NG)

细节辨认
3. In order to survive, a language needs to be spoken by more than 100 people. (Y)

细节辨认
2. Research on the subject of language extinction began in the 1990s. (N)

我们是如何知晓的？在过去的二三十年中，全球的语言学家就开始收集对比数据。如果他们发现一种语言只有少数人使用，而且没有人对语言的流传感到忧心，那么他们可以得出这种语言必定马上消亡。我们也可以得出同样的结论：一种语言如果只有不到 100 人使用，那么它一定不可能流传很久。1999 年的调查显示全世界 97％的语言只被 4％的人使用。

同义转述
5. The small community whose language is under threat can take measures to revitalize the language. (Y)

一旦一个社会开始认识到它们的语言处于危险之中，它就开始采取能够使其真正复活的措施。社会本身必须想要保护他们的语言，作为其中一部分的文化必须要尊敬少数民族的语言。需要有资金支持所开设的课程，使用的材料和授课的教师，需要语言学家继续从事把语言记录下来的基本工作。概括起来就是把语言归档——录音，分析，记录。人们必须能够阅读和

an increasingly computer-literate civilization.

But can we save a few thousand languages, just like that? Yes, if the will and funding were available. It is not cheap getting linguists into the field, training local analysts, supporting the community with language resources and teachers, compiling grammars and dictionaries, writing materials for use in schools.

It takes time, lots of it, to revitalize an endangered language. Conditions vary so much that it is difficult to generalize, but a figure of $900 million.

There are some famous cases which illustrate what can be done. Welsh, alone among the Celtic languages, is not only stooping its steady decline towards extinction but showing signs of real growth. Two language Acts protect the status of Welsh now, and its presence is increasingly in evidence wherever you travel in Wales.

On the other side of the world, Maori in New Zealand has been maintained by a system of so-called "language nests", first introduced in 1982. These are organizations which provide children under five with a domestic setting in which they are all intensively exposed to the language. The staffs are all Maori speakers from the local community. The hope is that the children will keep their Maori skills alive after leaving the nests, and that as they grow older they will in turn become role models to a new generation of young children. These are cases like this all over the world. And when the reviving language is associated with a degree of political autonomy, the growth can be especially striking, as shown by Faroese, spoken in the Faeroe Island, after the islanders received a measure of autonomy from Denmark.

In Switzerland, Romansch was facing a difficult situation, spoken in five very different dialects, with small and diminishing numbers, as young people left their community numbers in the German-speaking cities. The solution here was the creation in the 1980s of a unified written language for all these dialects. Romansch Grischun, as it is now called, has official status in parts of Switzerland, and is being increasingly used in spoken

书写。即使他们和他们的语言在计算机使用日益频繁的文明社会中有未来，人们也必须能读和写。

同义转述

6. A few thousand languages can be saved if enough funds are raised to do so. (**Y**)

要想做一些事情帮助恢复这些语言为时已晚，这些语言的使用者太少或太老，整个群体急于试图生存而无人关心这种语言。但是有一些语言还没有达到这种危险的程度。常常只有在一种语言濒危时，人们才会做一些事情来给他们注入新的生命，这叫做语言的复兴。

但是就像那样，我们能够挽救几千种语言吗？是的，如果可以利用人们的意愿和资金就可以。让语言学家研究该领域，培养当地的语言分析员，资助当地社区的语言资源和教师，编写语法和字典以及用于学校的书写材料，这些费用都不便宜。它需要花费很多的时间才能认识到这是一种将要消亡的语言。条件的多变以至于很难推断，但大约需 90 亿。

这里有一些能够说明我们所能做到的著名例子。威尔士语，单独存留在凯尔特语言中，不但稳固的停止走向消亡而且还显示出真正发展的迹象。两条语言法案都在保护着今天威尔士语的地位，无论你在威尔士的任何地方旅行，存在日益明显。

在世界的另一面，纽西兰岛的毛利语现已通过所谓的"语言巢"得以保存，该体制于 1982 年开始推行。所设立的机构为五岁以下的儿童提供集中授课的家庭环境。工作人员都是来自当地社区的毛利语使用者。该体制希望这些孩子离开"巢"后还能保留在生活中使用毛利语的技能，而且随着他们的成长顺势成为新一代儿童的行为榜样。世界各地都有类似这样的例子。当复兴的语言与某种程度的政治自治联系时，其增长是尤为显著的，如法罗语，是岛上居民从丹麦收回自治权后在法罗群岛使用的一门语言。在瑞士，拉丁罗曼语正在面临困难的处境，拥有五种不同的方言，由于居住在讲德语的城市里的年轻人丢弃了自己的民族身份而使使用的人数逐渐缩小。对此解决的办法是在 80

细节辨认

8. Romansch Grischun is a (an) _____ language in parts of Switzerland. (**official**)

form on radio and television.

A language can be brought back from the very brink of extinction. The Ainu language of Japan, after many years of neglect and repression, had reached a stage where there were only eight fluent speakers left, all elderly. However, new government policies brought fresh attitudes and a positive interest in survival. Several "semi-speakers"—people who become unwilling to speak Ainu because of the negative attitudes by Japanese speakers—were prompted to become active speakers again. This is fresh interest now and the language is more publicly available than it has been for years.

If good descriptions and materials are available, even extinct languages can be revived. Kaurna, from South Australia, is an example. This language had been extinct for about a century, but had been quite well documented. So, when a strong movement grew for its revival, it was possible to reconstruct it. The revised language is not the same as the original, of course. It lacks the range that the original had, and much of the old vocabulary. But it can nonetheless act as a badge of present-day identity for its people. And as long as people continue to value it as a true marker of their identity, and are prepared to keep using it, it will develop new functions and new vocabulary, as many other living language would do.

It is too soon to predict the future of these revived languages, but in some parts of the world they are attracting precisely the range of positive attitudes and grass roots support which are the preconditions for language survival. In such unexpected but heart-warming ways might we see the grand total of languages in the world increased.

细节辨认

9. The example of Ainu illustrates that a language can be saved _____.
(**from the very brink of extinction**)

年代后为这些方言建立了统一的文字。如今所称之的拉丁罗曼语在瑞士的部分地区已占有官方的地位，正在越来越多的通过广播和电视的形式发表。

细节辨认

7. An extinct language can never be revived no matter what you do about it. (N)

一种语言是能够从消亡的边缘恢复的。日本的虾夷语，在被多年的忽视和压抑后，已达到只有八个人能流利使用并且使用者全部都是老人的阶段。然而，在挽救方面新的政府政策带来了新的理念和积极的兴趣。很多母语半操用者——因为讲日本语的人的负面态度而不愿意说虾夷语的人——再度变得活跃。这是新的兴趣所在，而且这种语言比过去使用得更加公开化。

如果有详细的说明和材料可以使用，即使是消亡的语言也可以恢复。以澳洲南部的库纳语为例。这种语言已经消亡了一个世纪，但有很好的文件记录。所以，为了这种语言的复兴而开展有利的推广运动时，它就有可能重建。当然重建的语言会与原来的有所不同。缺乏原语言涉及的范围和絮聒古老的词汇。尽管如此仍可以作为一种标志而被今天的人们认同。只要人们能够把它当作身份的真正标识去珍惜，而且准备长久的使用，开发新的功能和词汇，那么其他存活的语言也会随之如此。

10. The preconditions for a language to survive is the people's _____.
(**positive attitudes and grass roots support**)

预测这些已经恢复的语言的前景还为时过早，但在世界的很多地方他们恰恰吸引着人们广泛的积极想法和作为语言生存先决条件的基础支持。在这种出人意料却让人感到温暖的方式下，我们可以看见世界语言的累积增长。

1.【题意】语言消亡的速度已经增加。

【解析】本句的关键词是 extinct。据此在第一段找到了 But what is happening today is extraordinary, judged by the standards of the past. It is language extinction on a large scale. 一句说用过去的标准作参照，一句说在大规模消亡。因此判断为 YES。

2.【题意】对语言消亡问题的研究是 20 世纪 90 年代才开始的。

【解析】本句的关键词是 in the 1990s。因为数字是最容易查找的，可以在第二段最后一句话找到，但该句并不是研究的开始，应根据此段的开头，How do we know? 及随后的答案为我们提供判断的依据。In the course of the past two or three decades, linguists all over the world have been gathering comparative data. 在过去 20－30 年间的全球的语言学家就开始收集对比数据，可见研究不是在 1990 年开始的。故此题选 NO。

3.【题意】一门语言要生存下来需要 100 以上的人说这种语言。

【解析】此句的关键词可确定为数字 100。在第二段，可以找到句子 And we have to draw the same conclusion if a language has less than 100 speakers。故此句选 YES。

4.【题意】世界上有些国家的语言比另一些国家的语言消亡得快。

【解析】此句关键词可确定为 language extinct。可以去第二段查找，第二段主要讲语言消亡的现象，但未提及有些国家的语言比另一些国家的语言消亡得快。故此题选 NG。

5.【题意】语言受到威胁的地区应当采取措施复活这一语言。

【解析】本句的关键词可以是 take measures 与 revitalize。在第三段出现了 revitalize 这个词并对其作了解释，下一段的第一句话与题目所述的意思相同，故此题选 YES。

6.【题意】如果筹集到足够的资金就可以挽救几千种语言。

【解析】本句的关键词是 a few thousand languages 与 funds。数字始终作为关键词的首选，可以在第五段找到。故此题选 YES。

7.【题意】不管采取什么行动，一种消亡的语言是再也不能复活的。

【解析】本句的关键词是 extinct 与 revive。revive 第一次出现在第八段，但相关的地方找不到 extinct，第二次出现在第 11 段，且两词均能找到，可以确定其出处，仔细阅读后发现与题目相反，故此题选 NO。

8.【题意】Romansch Grischun 在瑞士的部分地区是_____语言。

【解析】本句的关键词是 Romansch Grischun，专有名词在大多数的情况下都可确定为关键词，在第九段可以找到答案，此空填 official。

9.【题意】Ainu 的例子说明语言可以从_____拯救出来。

【解析】同上题一样，确定为关键词 Ainu，可在第十段找到答案，该段第一段是段落主题句，题目的意义与该句基本一致。答案是 from the very brink of extinction。

10.【题意】一门语言得以生存的前提条件是人民的_____.

【解析】本句的关键词是 preconditions，可以在最后一段找到相关答案，故此空填 positive attitudes and grass roots support。

Part Ⅲ   Listening   Comprehension
Section A

11. M: I'd like to speak to Mr. Jones, please.

   W: Sorry, sir. But Mr. Jones isn't here any more. Mr. Williams is in charge now.

   Q: What can we infer from the conversation?

   【答案及解析】 推理判断题。对话中男士说他想找 Mr. Jones，但女士说 Mr. Jones 已不在这儿，现在是 Mr.

Williams 负责。由此可以推论出 Mr. Jones 以前是这儿的负责人，故答案 C 正确。

12. M: Wow, there's a great deal of work for us to do.

W: Oh, it isn't as bad as it looks. After all, the greater part of it has already been done.

Q: What does the woman say about the work?

【答案及解析】信息明示题。对话中男士说还有一大堆工作等着去做，女士则说事情还不是那么糟糕，毕竟，工作的大部分已经做完了。句子 "…the greater part of it has already been done." 给了我们明确的信息，变被动为主动，意思即为 "they've finished more than half of it."，故选项 D 正确。

13. W: Can't you knock on the door before you enter my office next time?

M: Sorry, Madam. It's just that I'm in such a hurry.

Q: How did the woman feel when she was speaking to the man?

【答案及解析】态度题。对话中女士对男士说：下一次你进我的办公室之前能不能先敲一下门？男士马上说：对不起，只是因为我太匆忙了。问题问的是关于女士的态度，从女士的婉转建议中 "can't you knock on the door…?" 可推论出她有些生气，所以选 B 是正确的。

14. W: Jane told me she would fly to Paris sometime this week.

M: Well, I saw her a minute ago at the supermarket.

Q: What can we conclude from the conversation?

【答案及解析】推理判断题。对话中女士说 Jane 曾告诉过她要在这个星期的某个时间乘飞机到巴黎，男士则说一分钟之前他还在超市里见过她。由此，选项中的"撒谎"与"不撒谎"并不确定，但是很明显 Jane 今天不在巴黎。

15. W: Bob thinks you shouldn't use your good knife to fix that.

M: Tell him it's not his knife.

Q: What does the man imply?

【答案及解析】推理判断题。对话中女士对男士说：Bob 认为你不该用你那么好的小刀去修理那种东西。但男士说：告诉他这不是他的小刀。从男士的说话中可以推断出，男士认为 Bob 应该管他自己的事，言外之意为 Bob 是多管闲事。选项中词组 "mind one's business" 是"管闲事"之意。

16. M: How long will the party last? I've got a meeting to attend at 4 pm.

W: You'll be all right. The host will have an appointment at 3 pm.

Q: What do we learn about the man?

【答案及解析】推理判断题。对话中男士问女士：此次聚会要持续多长时间？四点钟，我还要去参加一个会议。女士则说：可以的，主人三点钟有个约会。由此可以推断出，男士不会因此错过了开会。

17. M: Why do you look so worried? Only one has finished ahead of you.

W: I've promised my Mom that I'd be the first.

Q: Why is the woman worried?

【答案及解析】因果关系题。对话中男士问女士：为什么你显得很焦虑？在你前面完成的就只有一个人呀。女士则说：我已向我妈允诺过要拿第一的。由此可以知道，她焦虑是因为她不能实现她的诺言。

18. W: My friend talked to me on the phone for two hours last night!

M: Is it toll free?

Q: What does the man imply?

【答案及解析】推理判断题。对话中女士说：昨晚朋友在电话里和我聊了两个小时。男士则问：电话费免费吗？该句中 "toll" 意指长途电话费，"free" 是 "免费" 的意思。由此我们可以推论出，男士认为他们通电话的时间太长，得花很多电话费。

Now you will hear two long conversations.

Conversation One

Joyce: Dad!

Father: Yes? What's the matter, Joyce?

Joyce: I'm wondering if I should buy a pair of tennis shoes. I'm going to join the tennis club in school.

Father: Why not? It's good that you finally play sports.

Joyce: But I'd like to have Adidas.

Father: Adidas? It's expensive. It's for the Chicago Bulls!

Joyce: No. All the guys in the school tennis team are wearing Adidas, boys, as well as girls.

Father: But none of us has ever had Adidas and we used to play quite OK.

Joyce: Here, Dad, is an ad about Adidas. Can I read it to you?

Father: Go ahead.

Joyce: "Over forty years ago, Adidas gave birth to a new idea in sports shoes. And the people who wear our shoes have been running and winning ever since. In fact, Adidas has helped them set over 400 world records in track and field alone."

Father: Nonsense! The players have to go through a lot of hard training and practice. It's nothing to do with the shoes. They may be comfortable, but...

Joyce: You're right, Dad. The ad goes on to say "You are born to run. And we were born to HELP YOU DO IT BETTER."

Father: Hmm. It may be good for running, but you don't run.

Joyce: Listen, "... Maybe that's why more and more football, soccer, basketball and tennis", see? "TENNIS players are turning to Adidas. They know that, whatever their game, they can rely on Adidas workmanship and quality in every product we make."

Father: OK, OK, dear. I know Adidas is good. But how much is a pair of your size?

Joyce: You don't have to worry about that, Dad. I've saved some money since last Christmas. I just want to hear your opinion.

Father: That's good. I have been wanting to have a pair of Adidas sneakers myself.

Questions 19 to 22 are based on the conversation you have just heard.

19. What is Joyce doing?

【解析】主旨大意题。整篇对话的重点是父女俩讨论 Adidas 运动鞋的优劣，因此正确答案是 C。

20. What does her father think about Adidas shoes?

【解析】事实细节题。当女儿提到想要买 Adidas 的网球鞋的时候，父亲的反应是 It's expensive；而当女儿利用一则广告作为理由时，父亲又说 They may be comfortable, but...由此可见父亲对 Adidas 运动鞋的态度是：可能会很舒适，但是太贵了。因此正确答案是 C。

21. What doesn't the father say in the conversation?

【解析】事实细节题。对话中确实提到了 C 项的内容，但是是从女儿的口中说出的，而不是父亲，其他三项均是父亲谈话的内容，因此正确答案是 C。

22. Why does the father object to Joyce's idea of buying Adidas?

【解析】事实细节题。当女儿一提到想要买 Adidas 运动鞋的时候，父亲的反应就是 It's expensive，而最后女儿说不用父亲掏钱，自己去年圣诞节开始已经攒够了鞋钱时，父亲的反应是 That's good，由父亲前后态度的转变可以推断出，父亲之所以反对 Joyce 买 Adidas 运动鞋，是因为他觉得鞋太贵了，他得花不少钱，故选 A。

Conversation Two

W: Is there anything I can help you with?

M: Yes, I was wondering if you could help me find some travel guides for this city? And maybe a road map, too.

W: We have a variety of books on that subject. Some are quite general while others are more specialized and specific.

M: I guess the generalized ones would do. I'm only staying here for a couple of days, but I want something that can give me more or less the feel of the city.

W: In that case, maybe this book will do. It has a lot of pictures and is easy to read. There are also road maps in it.

M: That's perfect. I'll take this one.

Questions 23 to 25 are based on the conversation you have just heard.

23．What might be the relationship between the two speakers?

【解析】推理判断题。女士问男士需要什么，男士说想买导游手册和地图，女士开始向男士介绍并推荐导游手册。据此判断，显然这是在书店里，两人是店主与顾客的关系。

24．What do we learn about the man from the conversation?

【解析】推理判断题，需要深刻了解对话的内容。在对话中男士谈到他要在这个城市停留几天，所以要买导游手册和地图，以便对这个城市有所了解，由此判断他对这个城市不熟悉。

25．Which kind of travel guides is the man most likely to buy?

【解析】事实细节题。女店员推荐了一本简单易读，并配有图片和地图的书，男士说好极了，I'll take it. (就买这本了。)

Section B

Passage One

A friend of mine told me that when he was a young man, he went to work as a teacher in one of the states of India. One day, he received an invitation to dinner at the ruler's palace. Very pleased, he went to tell his colleagues. They laughed, and told him the meaning of the invitation. They had all been invited, and each person who was invited had to bring with him a certain number of silver and gold coins. The number of coins varied according to the person's position in the service of the government. My friend's was not high, so he did not have much to pay. Each person bowed before the ruler, his gold went onto one hip, his silver went onto another hip. And in this way he paid his income tax for the year. This was a simple way of collecting income tax. The tax on property was also collected simply. The ruler gave a man the power to collect a tax from each owner of land or property in a certain area, if this man promised to pay the ruler a certain amount of money. Of course, the tax collector managed to collect more money than he paid to the ruler. The difference between sum of money he collected and sum of money he gave to the ruler was his profit.

Questions 26 to 28 are based on the passage you have just heard.

【内容概要】这篇文章主要讲述说话者朋友的一次经历，他在印度的时候被统治者邀请赴宴，但是要求拿和自己薪水相对应的钱，最后揭示这是统治者征收个人所得税的一个途径。

26．What do we know about the speaker's friend?

【解析】细节题。根据开头第一句话…he went to work as a teacher in one of the states of India. 判断正确答案是 D) He was once a school teacher in India.

27．What was the real purpose of the ruler's invitation?

【解析】细节题。文中有一句话 This was a simple way of collecting income tax. 可知答案是 B) To declare new ways of collecting tax.

28. What does the passage say about the tax collectors?

【解析】推断题。短文在最后说…if this man promised to pay the ruler a certain amount of money. Of course, the tax collector managed to collect more money than he paid to the ruler. 如果一个人允诺给统治者一定量的钱，集税者设法收集的比所要的数量多，故选 C) They tried to collect more money than the ruler asked for.

Passage Two

Around the year 1000 A.D., some people from northwest India began to travel westward. Nobody knows why. After leaving their homes, they did not settle down again, but spent their lives moving from one place to another, their later generations are called the Romany people, or Gypsies. There's Gypsies all over the world, and many of them are still traveling with no fixed homes. There are about 8,000,000 of them, including 3,000,000 in Eastern Europe. Gypsies sometimes have a hard time in the countries where they travel, because they are different. People may be afraid of them, look down upon them, or think that they are criminals. The Nazies treated the Gypsies cruelly, like the Jews, and nobody knows how many of them died in Hitler's death camps. Gypsies have their own language Romany. They liked music and dancing. And they often work in fairs and traveling shows. Traveling is very important to them, and many Gypsies are unhappy if they have to stay in one place. Because of this, it is difficult for Gypsy children to go to school, and Gypsies are often unable to read and write. In some places, the education authorities tried to arrange special traveling schools for Gypsy children, so that they can get the same education as other children.

Questions 29 to 31 are based on the passage you have just heard.

【内容概要】本文主要是对 Gypsies 迁徙情况的介绍，并且也谈到了他们在迁徙过程中遇到的一些问题，尤其是孩子的就学问题，也提到了针对这一情况的对策。

29. Why did the ancestors of Gypsies leave their home?

【解析】细节题。文中开头 Around the year 1000 A.D., some people from northwest India began to travel westward. Nobody knows why. 在公元 1000 年，有些人从印度西北部向西迁徙，无人知道原因。可知答案是 B) The reasons are unknown.

30. What is the attitude of some people toward Gypsies?

【解析】细节题。People may be afraid of them, look down upon them, or think that they are criminals.人们可能害怕他们，瞧不起他们，或认为他们是罪犯。故选 A) They are unfriendly to Gypsies.

31. What measure has been taken to help Gypsy children?

【解析】推断题。文中最后 In some places, the education authorities tried to arrange special traveling schools for Gypsy children, so that they can get the same education as other children. 在一些地方，教育机构安排了特殊的旅游学校给 Gypsy 学生们，判断答案是 C) Special schools have been set up for them.

Passage Three

As the car industry develops, traffic accidents have become as familiar as the common code. Yet, their cause and control remain a serious problem that has multiple causes. As the very least, it is a problem that involves three factors: the driver, the vehicle, and the roadway. If all drivers exercise good judgment at all times, there would be few accidents. But that is rather like saying that if all people are honest, there would be no crime. Improved design has helped make highways much safer. But the type of accidents continued to rise because of human failure and an enormous increase in the numbers of automobiles on the road. Attention is now turning increasingly to the third factor of the accident, the car itself. Since people assume that the accidents are bound to occur, they want to know how cars can be built to protect the drivers.

【内容概要】本文主要是对交通事故原因的分析。总的来说有三方面的因素：司机、高速公路、汽车，并针对各个因素的改进办法进行了探讨。

32. What does the speaker think of the causes of automobile accidents?

【解析】同义替换题。文中开头第二句话 "…their cause and control remain a serious problem that has multiple causes." (交通事故)的原因和控制归结于复杂的因素。"Multiple" 和 D) The causes are very complicated.中的 complicated 同义。

33. What measure has been taken to reduce car accidents?

【解析】推理题。文中 "Improved design has helped make highways much safer." 改良的设计使高速公路更加安全了，可知答案是 A) Improved highway design. 改良的高速公路设计。

34. What remains an important factor for the rising number of road accidents?

【解析】推理题。"But the type of accidents continued to rise because of human failure …" 因为人类的失误使事故类型还在增加推断应该是指 B) Drivers' errors. 司机的错误。

35. What is the focus of people's attentions today according to the passage?

【解析】推理题。"Attention is now turning increasingly to the third factor of the accident, the car itself." 注意正逐渐转向事故的第三个因素：汽车本身。由此可知正确答案是 D) Designing better cars. 设计更好的汽车。

Section C

Russia is the largest economic power that is not a member of the World Trade Organization. But that may change. Last Friday, the European Union said it would support Russia's (36) effort to become a WTO member.

Representatives of the European Union met with Russian (37) officials in Moscow. They signed a trade agreement that took six years to (38) negotiate.

Russia called the trade agreement (39) balanced. It agreed to slowly increase fuel prices within the country. It also agreed to permit (40) competition in its communications industry and to remove some barriers to trade.

In (41) exchange for European support to join the WTO, Russian President Putin said that Russia would speed up the (42) process to approve the Kyoto Protocol, and international (43) environmental agreement to reduce the production of harmful industrial gases.(44) These "greenhouse gases" trap heat in the atmosphere and, are blamed for changing the world's climate.

Russia had signed the Kyoto Protocol, but has not yet approved it. The agreement takes effect when it has been approved by nations that produce at least 55 percent of the world's greenhouse gases. (45) But currently, nations producing, only 44 percent have approved the Protocol. Russia produces about 17 percent of the world's greenhouse gases. The United States, the world's biggest producer, withdrew from the Kyoto Protocol after President Bush took office in 2001. So Russia's approval is required to put the Kyoto Protocol into effect.(46) To join WTO, a country must reach trade agreements with many trading countries that are also WTO members , Russia must still reach agreements with China, Japan, South Korea besides the United States.

Part IV   Reading Comprehension (Reading in Depth)
Section A

【短文大意】随着战争波及到世界很多角落，孩子们也被残酷地卷入了冲突。然而，在阿富汗、波斯尼亚、哥伦比亚，很多孩子参加了和平教育项目。孩子们在学会解决纷争之后做起了和平缔造者。哥伦比亚儿童和平行动甚至被提名为 1998 年的诺贝尔和平奖。在哥伦比亚，孩子们作为和平缔造者，研究人权和贫穷问题，最后在首都波哥大与其他五所学校成立了一个小组，被称作"和平学校"。

课堂上给学生机会用合作、和平行为替代愤怒、暴力行为。正是课堂上对每个人的关心和尊敬鼓励孩子们向前迈进一步，成为和平缔造者。幸运的是，教育者们在网络上找到很多资料，对推动孩子沿着和平的道路前进非常有用。1992年成立的"青年和平缔造者俱乐部"设立了网站为教师提供教学资源和开启善意行动的信息。世界儿童慈善中心呼吁重视儿童权利，帮助战争受害儿童。在班级里创建和平缔造者俱乐部是值得称颂的，它可以扩散到其他班级，最终将影响整个学校的文化氛围。

47.【答案及解析】K) 根据句意，空格处缺少名词，意为"活动，项目"。H) image 意为"影像，图像"，有的考生对此单词较陌生；victims 意为"受害者"，均与句意不符。

48.【答案及解析】M) 空格位于定冠词 the 之后，介词 of 之前，因此缺少名词，role 能够与 take on the …of 搭配。A) acting 意思接近句意，但是不能和 take on the…of 搭配。

49.【答案及解析】A) 空格句谓语动词为 studied，因此 as peacemakers 作主语 Groups of children 的定语，意为"是，作为"。此为现在分词作定语，选项中的另一个现在分词为 B) assuming 意为"假设"，不符合句意，可以排除。

50.【答案及解析】J) 空格处缺少谓语动词，直接宾语为 opportunities，因此需要及物动词。答案只能选 offer。

51.【答案及解析】D) 空格处单词作 ones 的定语，与 peaceful 并列，意义应该与 angry 相反。cooperative 意为"合作的"。C) comprehensive 意为"理解的，广泛的"，意义不符。

52.【答案及解析】G) 空格位于 take a step 之后，介词 toward 之前，需要副词。孩子需要向前迈进一步成为和平缔造者。take a step forward "向前迈进一步"。L) respectively 是副词，意为"分别地"，意义不符。

53.【答案及解析】F) 空格位于 be 动词与表语 useful 之间，因此需要副词修饰 useful。especially 符合题意。

54.【答案及解析】I) 空格处需要名词，意义与 resources 或者 teachers 并列。image 为名词，看似与 resources 或者 teachers 并列，但意义不如 information 恰当。

55.【答案及解析】O) 空格位于定冠词 the 之后，介词 of 之前，因此缺少名词，要帮助的应该是孩子——战争的受害者。

56.【答案及解析】E) 空格位于定冠词 the 之后，名词 school 之前，缺少定语。根据文章意思，先在一个班级创建和平缔造者俱乐部，然后扩展到其他班，最后影响到整个学校的文化氛围。entire "整个"符合题意。

Section B

Passage one

Drunken driving, sometimes called America's socially accepted form of murder, has become a national *epidemic*（流行病）. Every hour of every day about three Americans on average are killed by drunken drivers, adding up to an incredible 350,000 over the past decade.

A drunken driver is usually defined as one with a 0.10 blood alcohol content or roughly three beers, glasses of wine or shots of whisky drunk within two hours. Heavy drinking used to be an acceptable part of the American alcohol image and judges were *lenient*（宽容的) in most courts, but the drunken slaughter has recently caused so many well-publicized tragedies,

57. Which of the following best concludes the main idea of the passage?

A) Drunken driving has caused numerous fatalities in the United States.

B) It's recommendable to prohibit alcohol drinking around the United States.

主旨题 C) The American society is trying hard to prevent drunken driving.

D) Drunken driving has become a national epidemic in the United States.

58. Which of the following four drivers can be defined as an illegal driver?

31

especially involving young children, that public opinion is no longer so tolerant.

Twenty states have raised the legal drinking age to 21, reversing a trend in the 1960's to reduce it to 18. After New Jersey lowered it to 18, the number of people killed by 18-to-20-year-old drivers more thandoubled, so the state recently upped it back to 21.

Reformers, however, fear raising the drinkingage will have little effect unless accompanied by educational programs to help young people to develop "responsible attitudes" about drinking and teach them to resist peer pressure to drink.

Tough new laws have led to increased arrests and tests and in many areas already, to a marked decline in fatalities. Some states are also penalizing bars for serving customers too many drinks.

As the fatalities continue to occur daily in every state, some Americans are even beginning to speak well of the 13 years' national prohibition of alcohol that began in 1919, which President Hoover called the "noble experiment". They forget that legal prohibition didn't stop drinking, but encouraged political corruption and organized crime. As with the booming drug trade generally, there is no easy solution.

细节推断

A) A sixteen-year-old boy who drank a glass of wine three hours ago.

B) An old lady who took four shots of whisky in yesterday's party.

C) A policeman who likes alcohol very much.

D) A pregnant woman who drank a beer an hour ago.

59. In reformers' opinion, _____is the most effective way to stop youngsters from drinking alcohol.

A) raising the legal drinking age from 18 to 21

同义转述

B) forcing teenagers to obey disciplines

C) developing young people's sense of responsibility

D) pressing teenagers to take soft drinks

60. The rule that only people above 21 years of age can drink_____.

A) is a new law promoted by the twenty states

B) had been once adopted before the 1960's

C) has been enforced since the prohibition of alcohol

D) will be carried out all over the country

61. What is the author's attitude toward all the laws against drunken driving?

观点态度

A) Optimistic.

B) Pessimistic.

C) Indifferent.

D) Ironic

【短文大意】酒后驾驶，有时被称为美国社会公认的凶杀方式，已经在全国流行。平均每天每小时大约有3个美国人被酒后驾驶者撞死，在过去的10年间死亡总数惊人地达到35万人。

酒后驾驶通常指驾车者血液中含有0.10的酒精，或大约相当于在两个小时内喝了3支啤酒、几杯葡萄酒或威士忌。酗酒以前在美国是一种可以接受的饮酒习惯，而且大多数法院的法官对此也很宽容。但是醉酒杀手最近造成了这么多的引起广泛注意的惨剧，特别是累及到很多幼童，公众意见已经不再那么宽容。

有20个州已经把允许喝酒的法定年龄提高到21岁，扭转了20世纪60年代将年龄降至18岁的倾向。自从新泽西州把年龄降至18岁之后，被18－20岁驾车者撞死的人数增加了一倍以上，因此该州最近又将法定年龄提高到了21岁。

然而，改革者担心提高饮酒法定年龄作用甚微，除非辅以教育课程以帮助年轻人养成对饮酒"负责的态度"并教导他们抵制同龄人喝酒的压力。

新颁布的强硬法律在许多地区已使被逮捕和被测试的人数增加，也使死亡人数明显下降。一些州也开始处罚向顾客售酒过多的酒吧。

由于死亡事件每天在各个州仍不断发生，一些美国人甚至开始讲起始于1919年的全国13年禁酒的好处，那时胡佛总统称之为"高尚的试验"。他们忘了禁酒令并没有阻止饮酒，却促使了政治上的腐败和有组织犯罪的发生。随着毒品交易的普遍激增，看来没有轻而易举的解决办法。

57. 【题目译文】下列哪个选项最好地概括了文章的中心思想？

【答案及解析】C) 本文重点讨论的是美国政府和民众对酒后驾车这个现象的态度以及采取的一些措施，只有选项C)较为全面地概括了大意。

58. 【题目译文】下列四位司机中哪一位能被定义为非法驾驶者？

【答案及解析】A) 第2段第1句定义了什么是酒后驾驶，第3段第1句指出18岁以下的少年不能饮酒，根据这两个条件即可判断选项A)为正确答案。

59. 【题目译文】在改革者看来，_____是阻止青少年喝酒的最有效的方法。

【答案及解析】C) 由 Reformers, … unless … help young people to develop "responsible attitudes" …可以知道，选项C)的内容最符合题意。选项A)是一些州的做法。选项B)、D)与文章无关。

60. 【题目译文】只有年龄超过21岁的人才能喝酒的规定_____。

【答案及解析】B) 该句中的 reversing 引出的分词结构暗示60年代之前，合法的喝酒年龄曾为21岁，在60年代降为18岁，现在有20个州又重新把该年龄升回到21岁。由此可见，合法喝酒年龄为21岁的这个规定并不是什么新发明，从而可以排除选项A)，确定选项B)为正确答案。

61. 【题目译文】对于抵制酒后驾驶的法律规定作者的态度是怎样的？

【答案及解析】B) 本文最后一句中作者表明了对政府能否解决这个问题的忧虑: there is no easy solution, 故选 B)。作者在全文的叙述中表现出了对 drunken driving 这一社会问题的极大关注，因此选项C)是错误的；观点态度题往往需要在通读全文的基础上做出判断，但首尾句通常也透露了作者的一些感情色彩。

Passage Two

The economic effects are easy to see. Since 1978, some 43 million jobs have been lost largely to forms of technology —either to robotics directly or to computers that are doing what they are supposed to be doing, being labor-saving devices. Today, there is no such thing as a lifetime job; there is no such thing as a career for most people anymore. The jobs that are not done away with are being deskilled, or they are disposable jobs. Even for those jobs that many of you may feel secure with, there are people who are working on what are called "expert systems" to be able to take jobs away from doctors and judges and lawyers. The machine is capable of shredding（切碎）these jobs as well.

But it's not just the jobs. The economy of jobs and services is trivial compared to the "Nintendo capitalism" that now operates in the world. Four trillion dollars a day is shuffled（流通）around the earth. The inevitable result of a Nintendo economy — pulling itself apart, losing jobs, insecure—is the shriveling（萎缩）of the society in which it exists. What we have is an apartheid（种族隔离的）society, with growing gaps between the rich and poor, and the rich spending a lot of time protecting themselves from the effects of the poor.

62. According to the passage, information technology brings hazard to_____.

细节
推断

A) human society and natural environment.

B) natural environment and economy

C) domestic economy and human society

D) economy, human society and environment

63. From the context, we can infer that "Nintendo capitalism" means _____.

A) a capitalism that is prosperous

语义
理解

B) a capitalism that leads to destruction

C) a worship of capitalism

D) a worship of technology

64. The statement made by the author that what we have is an apartheid society because _____.

推理
判断

A) the rich are richer and the poor are poorer

B) the white are prejudiced against the black

C) a lot of jobs have been lost to high technology

D) the society is gradually shriveling

A further result of information technology—something that nobody seems to wish to pay much attention to—is the shredding everywhere of the natural world. Forget about the number of toxins that go into producing these computers, and the resources that go into producing them, such that 40,000 pounds of resources are necessary for a four-pound laptop. That's trivial compared with the direct effect that computers and the industrial system as a result have on the atmosphere and climate, the pollution of air and water.

The development in technology does not always bring human beings goods; there is bad news too. But most people are ignorant of the drawback of the new technology at first. In this century, however, the development in science and technology really aroused people's attention of the weak points.

But the technology has an even darker effect, because it is enabling us to conquer nature. Industrial society is waging a war of the techno-sphere against the biosphere. That is the Third World War. The bad news is that we are winning that war.

65. According to the third paragraph, what has caused the pollution of air and water?

A) The resources used to produce computers.

B) The four-pound laptop.

细节
推断 C) Computers and industrial system.

D) Shredding of the natural world.

66. When it comes to the term "the Third World War", the author tries to imply_____.

A) human conquering of nature.

推理
判断 B) information technology's destruction of natural environment.

C) technology's control over nature.

D) technology's conquering of human society and nature.

【短文大意】技术对经济的巨大影响不容忽视。自 1978 年以来，大约有 430 亿份工作让位于各种形式的高科技，要么直接由机器人来做，要么就由电脑来做那些本来就应该由节约劳动力的装置来做的事情。现在不再有能干一辈子的工作了，对大部分人来说也不存在"事业"这种说法了。有些工作虽没被淘汰，但技术含量降低了或变成了一次性的工作。即便是那些许多人认为很稳定的工作，也会有人开发"专家系统"软件以便抢走医生、法官和律师的饭碗，机器还会减少这些工作。

这不仅仅是就业问题。和目前存在的"任天堂资本主义"比较起来，就业和服务经济是微不足道的。每天四万亿美元在全球范围内移来移去。"任天堂经济"意味着支离破碎、失业和不安全。其导致了一个不可避免的后果，那就是它所处的社会在萎缩。我们的社会是一个种族隔离的社会，贫富之间的差距日益扩大，大多数时候，富人保护好自己的利益，漠视穷人的苦难。

信息技术的另一个更深远的影响是整个自然界的萎缩，人们似乎不愿意去注意这一点。为了生产计算机要耗费数量巨大的毒素和资源，一个重四磅的笔记本电脑要耗费四万磅之多的资源。忘记这些吧！这与电脑和受其影响的工业体系对大气和气候的直接影响——空气和水污染——比起来是微不足道的。

技术进步并不总为人类带来好处，也有它的负面影响，但是大多数人刚开始并不了解这种新技术的缺陷。然而本世纪科学技术的发展终于让人们注意到了它的弱点。不过技术革新有其更黑暗的一面，因为它使人类能够征服自然。工业社会对生物圈发起了一场技术领域的战争，这是第三次世界大战，不幸的消息是人类赢了。

62.【题目译文】根据文章，信息科技给_____带来了危险。

【答案及解析】D) 细节题。文章一至三段分别讲了信息技术对经济尤其是就业、社会和自然环境的影响，第四段为综述。故 D)最全面地概括了信息技术的负面影响，为正确答案。A)不全面，把对经济的影响漏掉了，故排除；B)同样也不全面，漏掉了对社会的影响，也排除；C)提到国内经济把经济范围缩小了，而文中提到的经济是全球范围内的，表述错误，故排除。

63.【题目译文】从上下文中，我们能够推断出"Nintendo capitalism"的含义为_____。

【答案及解析】B) 语意理解题。通过该短语后面的解释和说明，可猜出其意。从第二段 The inevitable result of a Nintendo economy—pulling itself apart, losing jobs, insecure—is the shriveling of the society in which it exists. 一句, 我们可断定该短语的含义是负面的, 可导致破坏或毁灭的, 故 B)符合题意, 为正确选项。

64.【题目译文】作者所说的我们所处的是一个种族隔离的社会因为_____。

【答案及解析】A) 推理判断题。答案见第二段最后一句 What we have is an apartheid(种族隔离的)society, with growing gaps between the rich and poor, and the rich spending a lot of time protecting themselves from the effects of the poor. 可见作者所说的种族隔离指的是富人和穷人的隔离, 故 A)正确。

65.【题目译文】根据第三段, 是什么导致了空气和水的污染?

【答案及解析】C) 细节题。答案见第三段最后一句: That's trivial compared with the direct effect that computers and the industrial system as a result have on the atmosphere and climate, the pollution of air and water. 从本句可以看出, 电脑及工业体系对大气和气候的直接影响就是空气和水污染。本题要求逆向思维进行反推。

66.【题目译文】当提到"第三次世界大战"这个词时, 作者试图暗示的意思是_____。

【答案及解析】B) 推理判断题。最后一段最后三句指出"工业社会对生物圈发起了一场技术领域的战争。这是第三次世界大战。不幸的消息是人类赢了。"可见强调的是技术革新对自然界的破坏性。作者此处以第三次世界大战为例, 指出这种破坏的严重性, 故 B)正确。

Part V   Error Correction

【短文大意】众所周知, 人们最早是把篝火上的枝形吊灯作为信号灯使用的。城堡的主人站在岩石错综的海岸边, 高举着涂满兽脂的吊灯, 警告路过的船只远离危险的海岸。

很快蜡烛成了信号灯的常用燃料。之后它被可以燃烧得更旺, 照明得更亮的油灯所取代。人们也尝试着使用煤油灯和汽油灯进行照明。至今仍有许多小灯塔在使用。但大部分的灯塔都装上了可以照射整个海面的电灯。

倘若古代的火把信号只是告诉人们"危险, 离开!", 那么今天的灯塔也同样可以通过信号确认船只。大部分的灯塔都有自己的专用信号。塔灯可能只是一眨一眨的——如同黑夜里巨大的萤火虫。或者像旋转灯, 一会儿红灯, 一会儿绿灯, 或只是白灯。然而无论信号如何, 它都是有规律地发送。这样航行在信号覆盖范围内的船只就不会对灯塔或灯塔的位置感到迷惑。

67. hangs→hung   从上下文看, 句中谓语动词应用一般过去时, 故将一般现在时 hangs 改为动词过去式 hung。

68. from→of   warn 不和 from 搭配。warn...of ...是固定搭配。

69. that→which   在从句中 that 作关系代词引导定语从句, 修饰 oil lamps。但是从上下文及从句前的逗号可以判断该定语从句为非限定性的, 不能用 that 引导。

70. lamps^→were   此处表达被动含义, 要用一般过去时的被动语态, 故要在 also tried 之前加 be 动词, 又因为主句为复数, 因此用 were。

71. sent→send   从上下文看, 此处应用一般现在时, 故应将 sent 改为 send。

72. it→itself   此处的 it 是指 lighthouse, 应改为 itself, 指代明确。

73. as→like   as 作介词用时, 意思是"作为, 如同", 而要表示"像……一样", 介词只能用 like。

74. revolved→revolving   此处 light 和 revolve 之间是主动关系, 不存在被动关系, 因此不能用过去分词, 而应改为现在分词形式。

75. however→whatever   whatever 修饰名词, 意思是"无论什么", however 修饰形容词、副词, 意思是"无论多么, 无论怎么"。此处修饰名词 signal, 故将 however 改为 whatever。

76. being→/   由上下文可知, 此处应为一般现在时的被动语态, 而不是一般现在进行时的被动语态。故将 being 删去。

Part VI   Translation

77. _____ (多亏你们的帮助), we accomplished the task in time.

【题目译文】多亏你们的帮助，我们及时完成了这项任务。

【答案及解析】Thanks to your help; thanks to 由于，多亏

78. _____ (这到底是谁告诉你的?) That is not true.

【题目译文】这到底是谁告诉你的？那不是真的。

【答案及解析】Who on earth told you that? on earth 究竟，到底。

79. _____ (假如遇到)fire, ring the alarm bell.

【题目译文】假如遇到火警，立即按警铃。

【答案及解析】In case of 假如，防备

80. It came to my ears _____ (偶然).

【题目译文】我偶然听到它。

【答案及解析】by chance 偶然，碰巧

81. I saw Tom and _____ (他一次拿六个盒子).

【题目译文】我看见汤姆了，他一次拿六个盒子。

【答案及解析】he carried six boxes at a time; at a time 同时；一次

大学英语4级考试新题型模拟试卷三答案及详解

参考答案

Part I Writing

My View on Going After Fashion

Nowadays some college students like to go after fashion . Boy students wear long hair and blue jeans and girl students wear short hair and mini skirts. When a new fashion appears on streets, these young people will just follow it without considering whether it is suitable to them or not. In their eyes, fashion is the symbol of the young generation.

However, some other students prefer to go their own way. They pay little attention to the common taste of fashion. They often seek the styles they really like. Though sometimes what they wear is quite different from what others do, they do look great and nice.

As to me, I don't like to go after fashion. Though fashion, in some way, may show a tendency of common taste, it depersonalizes people by making people everywhere look alike. I think only those who lack confidence will try to show their existence by these outside symbols. As college students, we should find ourselves and be ourselves. That will be the only kind of fashion that really counts.

Part II Reading Comprehension (Skimming and Scanning)

1. N	2. N	3. Y	4. Y	5. Y	6. N	7. Y

8. melting pot   9. 13 miles   10. map and routes

Part III Listening Comprehension
Section A
11. C	12. C	13. C	14. D	15. B	16. A	17. B	18. A
19. B	20. D	21. C	22. B	23. B	24. C	25. B	

Section B
26. C	27. D	28. A	29. A	30. A	31. B	32. B	33. D	34. D	35. C

Section C
36. persistence	37. career	38. ultimate	39. personal
40. seldom	41. separates	42. occasionally	43. defeat

44. Successful people learn from defeats, revise their strategy as needed and try again

45. Unsuccessful people try something one or two times and when it fails they give up

46. If you are persistent, you will almost inevitably succeed

Part IV Reading Comprehension (Reading in Depth)
Section A
47. K: small	48. L: possession	49. B: sprang	50. O: great	51. M: dangerous
52. I: equally	53. E: deserted	54. D: state	55. H: popularity	56. G: steep

Section B
57. D	58. A	59. C	60. A	61. B	62. C	63. B	64. D	65. B	66. A

Part V Error Correction
67. had →has	68. directly →indirectly	69. into →on	70. too →so
71. planet →planets	72. head →mind	73. little →much	74. Consider →Considering
75. They →删去	76. arriving →arriving at		

答案解析及录音原文

Part I　　　Writing

My View on Going After Fashion

Nowadays some college like to go after fashion . Boy students wear long hair and blue jeans and girl students wear short hair and mini skirts. When a new fashion appears on streets, these young people will just follow it without considering whether it is suitable to them or not. In their eyes, fashion is the symbol of the young generation.

However, some other students prefer to go their own way. They pay little attention to the common taste of fashion. They often seek the styles they really like. Though sometimes what they wear is quite different from what others do, they do look great and nice.

As to me, I don't like to go after fashion. Though fashion, in some way, may show a tendency of common taste, it depersonalizes people by making people everywhere look alike. I think only those who lack confidence will try to show their existence by these outside symbols. As college students, we should find ourselves. That will be the only kind of fashion that really counts.

【评论】本文是一篇议论文。开头直点主题，列举了当前学生追求时尚的种种现象。接着，话锋一转，谈到有些学生如何保持自己的个性。最后，作者表达了自己对待时尚的看法。这篇文章层次分明，条理清晰。

【常用句型】 通过列举各种现象，然后表达自己看法的作文经常用到以下句型：

第一段：直点主题，列举现象 Nowadays some people like to do…

第二段：话题转折，谈到另一种情况 However, some other people prefer to go their own way. They pay little attention to …

第三段：表达自己的观点 As to me, I don't like to do …

Part II　　　Reading Comprehension (Skimming and Scanning)

【文章及答案解析】 本篇文章向我们介绍了美国著名的城市—纽约，概述了纽约的全貌、"大苹果"名称的由来、著名的游览景点以及相关的公共配套设施，全方面地展示了纽约这个世界级大都市的风采。

New York

As a travel destination, New York has something to offer almost every visitor. Though tourism has dropped since September 11, 2001, there are still lots of reasons to visit what many consider the greatest city in the world.

City Overview

New York City (NYC) is located on the eastern Atlantic coast of the United States. It rests at the mouth of the Hudson River. The city is often referred to as a "city of island". Greater NYC is made up of five distinct

【文章大意】　　　纽约

作为旅行的目的地，纽约可以满足几乎每一位游客的需要。尽管 2001 年 "9·11 事件" 后旅游业有所下滑，但仍有很多缘由使得游客们想要参观这个被众人称之为世界上最大的城市。

主旨
大意

7．This passage is mainly about getting around New York City. (**Y**)

38

areas called boroughs. These boroughs include Manhattan, Brooklyn, Queens Island and the Bronx. The boroughs are separated from each other by various bodies of water and are connected by subways, bridges and tunnels.

When people refer to New York City, they are usually talking about Manhattan. Most of NYC's main attractions are located in this borough and the majority of visitors spend most of their vacation here.

A Short History of the Big Apple

No discussion about New York City would be complete without asking why New York is referred to as "the Big Apple". Like many things about New York, you'll probably get a different answer depending on who you ask. According to the Museum of the City of New York, it is believed that in the 1920s, a sportswriter overheard stable hands in New Orleans refer to New York City's racetracks as "the Big Apple". The phrase was most widely used by jazz musicians during the 1930s and 1940s. They adopted the term to refer to New York City, and especially Harlem, as the jazz capital of the world.

The Italian navigator Giovanni da Verrazano may have been the first European to explore the New York region in 1524. More than 80 years later, Englishman Henry Hudson sailed up the river that now bears his name. But it was Dutch settlements that truly started the city. In 1624, the town of New Amsterdam was established on lower Manhattan. Two years later, according to local legend, Dutchman Peter Minuit purchased the island of Manhattan from the local Native Americans for 60 guilders (about $24) worth of goods.

Few people realize that New York was briefly the U.S. capital from 1789 to 1790 and was the capital of New York State until 1797. By 1790, it was the largest U.S. city. In 1825, the opening of the Erie Canal, which linked New York with the Great Lakes, led to continued expansion.

城市概括

纽约市（NYC）位于美国东部哈得逊河口，濒临大西洋，经常被人们称作"岛城"。该城由曼哈顿、布朗克斯、布鲁克林、昆斯和里士满 5 个区域组成。彼此间由不同的水域分隔，并且有多条地铁、桥梁和河底隧道贯穿相连。

当人们提及纽约市时，他们通常会谈论曼哈顿区。它拥有纽约城的大部分景点，多数的旅游者都选择在这里度过他们假期中的大部分时间。

> 细节辨认
> 1. People usually use the name New York City for Brooklyn. (N)

大苹果的历史

谈及纽约难免会问道纽约为什么会叫做"大苹果"？就像纽约的很多事物一样，不同的人给予不同的回答。根据纽约市博物馆的资料记载，其源于 20 世纪 20 年代，一位体育新闻记者在新奥尔良无意中听到有人称纽约的赛马跑道为"大苹果"。之后的 30、40 年代里，这个词被广泛使用，尤其是作为世界爵士乐首都的黑人莱姆区的爵士乐者们。

> 细节辨认
> 2. There is a unanimous agreement as to why New York is referred to as "the Big Apple". (N)

1524 年意大利人弗拉赞诺作为第一个欧洲人来到河口地区。80 多年后，英国人哈得孙沿着后来以他的名字命名的河流上溯探险。但是真正建立起城市的是荷兰殖民者。1624 年他们在曼哈顿岛的下游建立了新阿姆斯特丹镇。据当地传说两年后，荷兰总督明奴特以价值大约 60 个荷兰盾（相当于 24 美元）的小物件从印第安人手中买下曼哈顿岛。

> 同意转述
> 3. New York was the U.S. capital from 1789 to 1790. (Y)

只有少数人知道纽约在 1789—1790 年曾是当时美国的临时首都，直到 1797 年成为纽约州的首府。纽约在 1790 年之前一直是美国最大的城市。1825 年，连接纽约和五大湖区的伊利运河建成通航，促进了城市的进一步扩展。

A charter was adopted in 1898 incorporating all five boroughs into Great New York. New York has always been and remains a city of immigrants. Patterns of immigration are integral to the city's history and landscape. Immigration, mainly from Europe, swelled the city's population in the late 19th and early 20th centuries. After World War II, many African-Americans from South, Puerto Ricans, and Latin Americans migrated to the city as well. Because of the variety of immigrant group, both historically and currently, New York is often referred as a true "melting pot".

1989 年通过的宪法将五个区全部纳入纽约。纽约一直都是一个移民城市。对于一个城市的历史和景观来说，移民的模式是整体性的。纽约的移民主要来自欧洲，19 世纪末到 20 世纪初人口开始膨胀。二战后，许多美国黑人也开始从南方、波多黎各和拉丁美洲迁移至此。由于移民群体的多样化，历史上和今天仍称纽约为"熔炉"。

8. Owing to the variety of immigrant groups, New York is often referred to as a true _____. (melting pot)

9. The length of New York City is _____. (13 miles)

Getting Around：By Foot

The absolute best way to get around New York, and the one you will probably be using most, is walking. Remember, the city is only 13 miles long. On a day with good weather, walking is a great option. The excitement of New York on foot is that you never know what interesting thing you will see as you head from one destination to another. If you're on a schedule, keep in mind that distances are not as close as they might seem and take into account the extra time it takes to stop at every street crosswalk. Getting from the easternmost side of Manhattan to the westernmost side can take quite a while.

漫步纽约

游览纽约最好也是最适用的方式就是步行。要知道，纽约市只有 13 英里长。在天气好的日子里，步行是最佳的选择。漫步纽约时最让人激动的是当你从一个地点到另一个地点时，你永远都不知道你将会看到什么样的有趣事情。如果按照行程表前行，应注意距离比看见的要远得多而且还应考虑在斑马线停留所花费的额外时间。从曼哈顿的最东部到最西部要花费很长时间。

The Subway

If you're looking to save some time, this is where one of the three excellent New York public transportation systems comes in handy. They are all run by the city's Metropolitan Transportation Authority. According to NYC &Company, the city's official visitor's bureau, the 714—mile New York City subway system has 468 stations serving 24 routes—more than any other system in the world. It operates 24 hours a day, is safe, and is used daily by more than 3.5 million people.

The main thing to remember when using the subway system is to make sure you get on the correct train. Uptown trains head north, downtown and Brooklyn borough trains head south. Express trains as opposed to local trains; do not make all the normal stops on the line.

地铁

如果你想节约时间，纽约最棒的三大公共交通体系中的任何一个都会给你提供便利。该系统由大都会捷运局管理。根据纽约旅游局、纽约市官方旅游署统计，纽约市的地铁系统全长 714 英里、468 个站台、24 条线路——超过世界上任何系统。全天 24 小时营运，安全有效，每天运送的人口超过 350 万。

同意转述

4. New York City subway system ranks first in the world in terms of the number of stations. (Y)

在搭乘地铁时，最主要的是要确定你所乘坐的火车是无误的。开往郊区的火车头向北，开往市区和布鲁克林区的火车头向南。与普通车不同的是，捷运不

Local trains make every stop. New Yorkers and tourists alike have hopped on the wrong train and ended up in an unknown area. If this happens, simply hop the next train back the way you arrived. To avoid these problems, when in doubt, always take a local train.

在任何的站点停留，而普通车每个站点都停。纽约人和游客一样，都会咒骂搭错的火车，然后在不知名的地方等待。若遇到这样情况，最简单的方法是坐下一辆火车按原路返回。通常在犹豫不决时，可选择普通车以避免类似事情的发生。

Buses &Taxis

The final two systems of public transportation are buses and taxis. Buses tend to be very slow because of New York traffic, but they can give you great views of the city streets. Buses run north and south as well as east and west. Just like the subway, the bus system has its own map and routes. Most free subway maps also include a bus map.

Taxis are usually quicker to navigate the city streets than buses. But be prepared to pay for that convenience. Taxis are expensive. A trip from the Upper West Side, for example, to the lower East Side can cost upwards of $12, not including the driver's tip. There is an automatic $2 charge on all cab rides and all taxi drivers expect some sort of gratuity.

公交车与出租车

三大公共交通系统的后两个体系就是公交车和出租车。由于纽约的交通流量大，公交车通常很慢，但你可以借此游览一下纽约的街景。公交车贯穿东西南北。和地铁一样，公交车也有自己的路线图。大多数免费的地铁图中也附有公交图。

10. The bus system is like the subway in that it has its own ___. (**map and routes**)

通常出租车在城市街道的行驶要比公交车快得多，但也要准备支付由此方便带来的费用。出租车费很不便宜，如从曼哈顿西区到东区不包括给司机的小费要 12 美元。每辆出租车上都有一个 2 美元的自动计费器而且所有的司机都希望能得到一点酬劳。

Main Attractions

Here is a small sapling of some of the main attractions NYC is best known for:

Statue of Liberty—Few New York sites are as awe inspiring as this one. A century ago, Lady Liberty held up her torch to welcome immigrants to America. Today, you can climb 354 steps to look out from her crown to see both the New York and New Jersey coasts

Times Square—New Yorkers call this intersection of Broadway and 42nd street the "Crossroad of the World". It is the most recognized intersection on earth—millions of people see it on television every New Year's Eve. Some people say it's the best place in New York to people-watch. At night, the illuminated signs in Times Square make an amazing light show.

Empire States Building—built in 1931, this skyscraper was the tallest in the world for half a century. You'll get a great view of the city from the art deco

主要的景点

下面简单介绍一下大家熟知的纽约著名景点：

自由女神——很少有像自由女神这样的景点能够引发人们的敬畏。一百年前，女神高举火把欢迎来到美国的移民者。今天，你可以爬上 354 级台阶从女神的王冠上俯瞰纽约和新泽西海岸。

时代广场——纽约人把百老汇和 42 街的交叉点称为"世界十字路口"。每年有数百万人在这个世界公认的十字路口等待除夕夜的新年倒数，据说这是纽约市的最佳位置。夜晚，可以欣赏到时代广场独特的招牌奇观。

同意转述

5. Times Square is the most recognized intersection on earth. (**Y**)

帝国大厦——修建于 1937 年，在过去的半个世纪中曾是世界上最高的建筑，你可以在带有装饰艺术风格的圆塔观景台上看城市的美景。

tower's observation desk.

Central Park—Who would have thought that a city filled with people, traffic and skyscrapers, could offer visitors such an incredible natural oasis? The park is full of rolling meadows, trees, water bodies and stone bridge. The best part? It's all free.

Metropolitan Museum of Art— If you see only one museum in New York City, the Met, as it is known, should be the one. The museum houses over two million works of art ranging from Egyptian to Medieval to 20th.

United Nations—You can't miss the 188 nations' flags high above First Avenue in front of the headquarters of this international organization. Tours take you through the Security Council and General Assembly Halls.

中央公园——不知是谁拥有这样的想法,在人群拥挤,车辆纵行,高楼林立的城市里为游客提供这种不可思议的自然绿洲。公园内有起伏的草地、茂密的树丛、大片的水域和石桥。最棒的莫过于公园是免费开放的。

同意转述 6. If you see only museum in New York City, you should visit the American Museum of Natural History. (N)

大都会艺术博物馆——如果在美国想参观一个博物馆,众所周知,大都会博物馆应是首选。博物馆藏有从古埃及到 20 世纪的 200 多万件艺术作品。

联合国——你不可以错过观看 188 个国家的旗帜高高飘扬在位于国际性组织总部前第一大街的场面。导游会带你参观安理会和联合国会议大厦。

1. 【题意】人们所说的纽约市往往指的是 Brooklyn(布鲁克林)。

【解析】在 City Overview 部分第一段指出纽约市分为五个区,而第二段首句明确提到 When people refer to New York City, they are usually talking about Manhattan. 即人们所说的纽约市往往指的是曼哈顿,而不是题干中的 Brooklyn。故本题应选的是 NO。

2. 【题意】关于纽约被称为 "the Big Apple" 的来历有统一的说法。

【解析】本题关键词 the Big Apple (专有名词以及具有特殊含义的词语往往是定位的信息词),将查找范围锁定在 A Short History of the Big Apple 部分。根据第二句中的 a different answer 可知,关于 the Big Apple 的来历众说纷纭,故此题选 NO。

3. 【题意】纽约在 1789-1790 年间曾是美国的首都。

【解析】从 from 1789 to 1790 可以判断本题与纽约的历史相关,于是在 A Short History of the Big Apple 部分寻找答案出处。第三段第一句提到了纽约作为美国首都的情况,时间是 1789 到 1790 年,所以题干表达的意思与此相符,故本题答案是 YES。

4. 【题意】在车站设置的数量方面,纽约市的地铁系统处于世界的首位。

【解析】根据主语 New York City subway system 将查找的范围锁定到 The Subway 部分,其中第一段指出 New York City subway system … more than any other ranks first 是一致的。故本题应选的是 YES。

5. 【题意】时代广场是公认的 "世界十字路口"。

【解析】题干的话题是时代广场,可以在相关的部分找到与时代广场有关的答案,其中提到 It is the most recognized intersection on earth,说明题干与原文相符,故本题应选的是 YES。

6. 【题意】纽约的自然历史博物馆是最值得参观的博物馆。

【解析】有关 Museum 的话题可以在 Main Attractions 部分的 Metropolitan Museum of Art 中查找,这一段指出 "If you see only one museum in New York City, the Met, as it is known, should be the one",因此题干的 the American Museum of Natural History 与文中不符,故本题应选的是 NO。

7. 【题意】本文主要介绍了纽约市的概况。

【解析】文章先对纽约进行了简单的介绍,讲述了发展史、城市规模、交通状况以及主要的景点,因此本题的陈述符合题干的意思，故本题应选 YES。

8．【题意】移民群体的多样性使得纽约被人们称为大熔炉。

　　【解析】原文在 A Short History of the Big Apple 部分最后一句提到，由于移民群体的多样性，纽约被人们称为"大熔炉"。because of 与题干的 owing to 同义。可见本题要填写的是 melting pot。

9．【题意】纽约市的长度是 13 英里。

　　【解析】题干问纽约市的长度，在 Getting Around New York On Foot 部分提到 "Remember, the city is only 13 miles long"。题干用 long 的名词形式 length，所以此处可填 13 miles。

10．【题意】同地铁一样，公共汽车也有自己的路线图。

　　【解析】根据题干的主语 bus system，可查看 Buses &Taxis 的相关内容。第一段提到 Just like the subway, the bus system has its own map and routes，题干对原文的句式进行了一定的改动，变成了一个主从复合句。由此可知本题该填写的答案为：map and routes。

Part Ⅲ　　Listening Comprehension

Section A

11．W: Good evening, sir. Here is the menu. Would you like to order now?

　　M: Yes, but I'm in a rush. Can I be served and out of here in half an hour?

　　Q: Where are the speakers?

　　【答案与解析】判断推理题。"Here is the menu. Would you like to order now?"是餐厅服务员接待客人的常用语，因此地点在餐厅。

12．W: How do you feel about the oral test?

　　M: I couldn't feel better about it. All of the questions are within my expectation.

　　Q: How does the man feel about the test?

　　【答案与解析】判断推理题。回答表明男士对口语考试感觉很好，因为所有的题目都在意料之中，因此选 C（自信）。

13．W: I spent $50 on this sofa. Do you think it is worthwhile?

　　M: Well, I think you've got a real bargain.

　　Q: What does the man think of the price?

　　【答案与解析】惯用法理解题。"you've got a bargain"是惯用法，意为"你买到了便宜货"，故选 C。

14．W: I heard the fire broke out at two in the morning.

　　M: That's right. And it took the fireman three hours to put it out.

　　Q: When was the fire put out?

　　【答案与解析】推理题。根据对话，凌晨两点着火，用了三个小时扑灭，因此扑灭的时间是五点钟。

15．M: Excuse me, I'd like to send a registration letter to Houston, Texas. How much is the postage?

　　W: Let me check. It's a dollar and 55 cents.

　　Q: What is the probable relationship between the two speakers?

　　【答案与解析】判断推理题。由对话可知，男士要寄一封挂号信，询问女士邮资是多少，因此应选 B。

16．W: John, there is a really good view out there.

　　M: Sorry, honey. I can't stop here. There's nowhere to park.

　　Q: What does the woman ask the man to do?

　　【答案与解析】判断推理题。女士说外面风景很好，男士却说抱歉此处不能停车。由此判断女士想找个地方停车。

17. W: Jack, I expected to see you at Sam's birthday party yesterday but you were absent.

M: I had to date with my girlfriend. I have been terribly busy these days. You know, she is complaining.

Q: What do we know about the man?

【答案与解析】细节题。男士的回答表明他没去晚会，而是和女朋友去约会了，因为他最近一直很忙而遭到了女友的抱怨。

18. M: I'm fed up with the noisy environment here. I suppose we should rent a house with better environment, but I don't see how we can afford it right now.

W: If only we hadn't bought the second-hand car!

Q: What does the woman mean?

【答案与解析】句式理解题。句式"If only..."通常接虚拟语气，意为"要是……该多好啊"，表示对现在或过去事实的一种虚拟假设。由此可知女士对买了二手车感到后悔。

Now you will hear two long conversations.

Conversation One

W: Good evening, Mr. Brown, and welcome to the program.

M: Thank you!

W: I wonder if you could tell listeners more about the work you do?

M: Certainly.

W: Mr. Brown, I gather that you work for a company developing software for reading electronic books?

M: That is so.

W: Do you think there is much future in that?

M: Well, people buy music online—so why not books?

W: You have to read electronic books on a machine, don't you?

M: That is so, although of course you might be able to print them out on a printer.

W: So why should people read a book on an expensive machine when they can buy a cheap copy and carry it around with them and read it whenever they like?

M: That is an interesting point. People need time to become aware of the value of e-books.

W: Is it true that at the Frankfurt Book Fair, in 2000, there was a prize for the best books published in electronic form?

M: Yes, that is true: the prize was worth $ 100,000.

W: Who put up the money for the prize?

M: Err, software companies such as Microsoft and Adobe.

W: I suppose they are trying to encourage publishers to get the e-book business off the ground.

M: That's probably true.

Questions 19 to 22 are based on the conversation you have just heard.

19. Where is this conversation taking place?

【解析】推理判断题。根据关键词 welcome to the program(欢迎来到我们的节目)和 listeners(听众)推断对话者在做广播节目。

20. What is Mr. Brown's attitude to electronic books?

【解析】推理判断题。对于女士就电子图书前景的询问，男士采取了一个反问法，他说："people buy music online—so why not books? 既然现在人们在网上购买音乐，那为什么不会购买图书呢？说话者反问的语气表达了

44

他的言外之意：电子图书有发展前景。

21．What is the woman's attitude to electronic books?

【解析】推理判断题。说话者的语气和语调是推理的重要依据。女士用了一个 why should …when …结构。指出当人们能买到便宜并且可随身携带的书进行随心所欲的阅读时，他们怎么会到昂贵的机器上读书呢？疑问句采用降调时表示置疑，否定，显然女子对电子图书怀有抵触情绪，不是很友好。

22．What was the prize for?

【解析】事实细节题。由女士问句: Is it true that at the Frankfurt Book Fair, in 2000, there was a prize for the best books published in electronic form? 可知此奖项是 "最佳的电子图书"。

Conversation Two

W: Jack, I understand your jazz band is going to play at the Student Center Ballroom. I just saw a poster advertising the event, and I called to tell you I'll be there.

M: Oh, thanks, but I'm not in that band any more. In fact, I'm not in a group at all right now.

W: That's too bad—you're such a talented musician. Why did you leave the group?

M: I just couldn't be a full-time student and play with the band every night. I missed a couple of courses during my mid-term exams, and I thought I'd better quit before the bandleader fired me.

W: Say, you know my friend Charlie, don't you? He plays saxophone and trumpet, and he and some of his friends are getting a band together. I bet they could use a good drummer.

M: I wouldn't have time for that, either.

W: Oh, I don't think they'll practice very often. They are students, too.

M: Do they plan to perform?

W: I don't know. Here's his number. You can get in touch with him.

Questions 23 to 25 are based on the conversation you have just heard.

23．Why did the man leave the band?

【解析】事实细节题。快速浏览已给的选项，可知本题是关于男士做某事的原因的题目。男士提到每晚排练占用了学习时间导致期中考试遇到了困难，所以他退出乐队是因为没有时间。

24．What role did the man probably play in the band?

【解析】事实细节题。女士提到一个朋友重组了乐队，继而提醒男士他们可能缺鼓手，因此推断男士可能是鼓手。

25．What did the woman suggest that the man do in order to contact her friend?

【解析】推理判断题。本题是关于要做的事情的题目，对话的最后部分，女士把乐队的电话给了男士并鼓励他加入: Here's his number. You can get in touch with him.

Section B

Passage one

Paul, a salesman from London, was driving past a sports car parked outside a supermarket when he saw it start to roll slowly down the hill. Inside the car were two young girls on the passenger seat — but no driver. Paul stopped quickly, jumped in front of the sports car and tried to stop it, pushing against the front of the car. Another man who was standing nearby got into the car and put on the handbrake, saving the girls from injury. It was at this point that Paul noticed his own car rolling slowly down the hill and going too fast for him to stop it. It crashed into a bus at the bottom of the hill and was so badly damaged that it had to be pulled away to a garage. As if this was not bad enough, Paul now found he had no one to blame. He was so busy chasing his car that he didn't get the name of the driver of the sports car

45

who just came out of the supermarket and drove away without realizing what had happened.

【内容概要】本文讲述了一个关于保罗的故事，一天保罗在超市外看见一辆没有司机、但车内有两位女士的车向山坡下滑去，他救了她们，自己的车却滑向山崖，严重损坏，但他却找不到人赔偿，他救的那辆车的司机在超市里，根本不知道发生了什么事情。

26. Which car was badly damaged?

【解析】 推断题。根据…Paul noticed his own car rolling slowly down the hill… and was so badly damaged…判断答案是 C) Paul's car.

27. Where was the driver of the sports car when the accident happened?

【解析】 推断题。根据…who just came out of the supermarket…判断，司机那时在超市里故选 D) In the supermarket.

28. Who did Paul think was to blame for the accident?

【解析】 推断题。根据文章后面的叙述，保罗找不到人追究责任，因为他不知道那个被他救了的车的司机的名字，因为在保罗追自己的车的时候，他从超市里出来把车开走了，故选 A) The driver of the sports car.

29. Who was injured in the accident?

【解析】 推断题。根据全文的叙述，只有保罗的车遭到了损坏，但是人员无伤亡，故选 A) Nobody.

Passage Two

My friend, Vernon Davies kept birds. One day he phoned and told me he was going away for a week. He asked me to feed the birds for him and said that he would leave the key to his front door in my mailbox. Unfortunately, I forgot all about the birds until the night before Vernon was going to return. What was worse, it was already dark when I arrived at his house. I soon found the key Vernon gave me could not unlock either the front door or the back door. I was getting desperate. I kept thinking of what Vernon would say when he came back. I was just going to give up when I noticed that one bedroom window was slightly open. I found a barrel and pushed it under the window. As the barrel was very heavy, I made a lot of noise. But in the end, I managed to climb up and open the window. I actually had one leg inside the bedroom when I suddenly realized that someone was shining a torch up at me. I looked down and saw a policeman and an old lady, one of Vernon's neighbours. "What are you doing up there?" said the policeman. Feeling like a complete fool, I replied, "I was just going to feed Mr. Davies's birds."

【内容概要】本文是讲述这样的故事：说话者的朋友外出，把他养的鸟交代给说话者照顾，但是他却忘记了，等他想起的时候去朋友家，找到朋友留给他的钥匙却打不开门，没办法他踩个大桶往屋子里爬，却见警察在下面，说话者只好解释说："我来喂鸟！"

30. Why couldn't the man open the door?

【解析】推断题。根据 I soon found the key Vernon gave me could not unlock either the front door or the back door. (我很快发现 Vernon 留给我的钥匙，前后门都打不开。)判断答案是 A)His friend gave him the wrong key.

31. Why did the man feel desperate?

【解析】 推断题。根据 I kept thinking of what Vernon would say when he came back. (我一直在想他回来会说什么)判断答案是 B) He was afraid of being blamed by his friend.

32. Why did the man feel like a fool?

【解析】 推断题。根据他的回答 I replied, "I was just going to feed Mr. Davies's birds." (我要喂戴维先生家的鸟)判断是 B) He knew the policeman wouldn't believe him. 警察不会相信他的话。

Passage Three

When Iraqi troops blew up hundreds of Kuwaiti oil well at the end of Gulf War, scientists feared environmental disaster. Would black powder in the smoke from the fires circles the globe and block out the sun? Many said "No way. Rain would wash the black powder from the atmosphere. But in America, air-sampling balloons have detected high concentrations of particles similar to those collected in Kuwait. Now the fires are out, scientists are turning their attentions to yet another threat: the oil that didn't catch fire. It has formed huge lakes in the Kuwaiti desert. They trap insects and birds, and poison a variety of other desert animals and plants. The only good news is that the oil lakes have not affected the underground water resources. So far, the oil has not been absorbed because of the hard sand just below the surface. Nothing, however, stops the oil from evaporating. The resulting poisonous gases are choking nearby residents. Officials are trying to organize a quick clean-up, but they are not sure how to do it. One possibility is to burn the oil and get those black-powder detectors ready.

【内容概要】本文是关于海湾战争之后,伊拉克军队炸开几百个科威特油井给全球环境所带来的危害的讨论。

33. What were the scientists worried about soon after the Gulf War?

【解析】细节题。文章开头 scientists feared environmental disaster. Would black powder in the smoke from the fires circles the globe and block out the sun? （科学家们害怕环境灾难，着火所放出的烟中的黑色粉尘会不会环绕地球，遮挡住太阳）判断答案是 D) The spread of the black powder from the fires.

34. What was the good news for scientists?

【解析】细节题。根据 The only good news is that the oil lakes have not affected the underground water resources. （唯一的好消息是石油湖没破坏地下水资源）判断答案是 D) The underground water resources have not been polluted.

35. What are the officials trying to do at the moment?

【解析】推断题。根据 Officials are trying to organize a quick clean-up （官方在组织大的清除）判断答案是 C) To remove the oil left in the desert.

Section C

The lack of (36)persistence is the reason most people fail in attaining their goals. Many organizational analysts and (37)career consultants consider persistence to be the (38)ultimate key to success at both the organizational and (39)personal level. Success (40)seldom comes easily on the first try. What (41)separates the successful from the unsuccessful is persistence. Successful people also fail (42)occasionally but they do not let their failures (43)defeat their spirit. (44)Successful people learn from defeats, revise their strategy as needed and try again. And again and again. Until they succeed. (45)Unsuccessful people try something one or two times and when it fails they give up, usually passing the blame on to someone or something else, and learn nothing from their experience other than perfecting their scapegoat techniques. Successful people expect periodic defeats, learn what went wrong and why, don't waste time looking for someone to blame, make necessary adjustments, and try again. (46)If you are persistent, you will almost inevitably succeed. If you are not persistent, you will almost certainly fail.

Part IV   Reading Comprehension (Reading in Depth)
Section A

【短文大意】当被问到最喜欢的城市的名字时,许多美国人都会选择旧金山。旧金山最初是一个以西班牙文命名的靠近海湾的小村落。美国于 1846 年在与墨西哥的战争中将其占领,当时城镇的规模仅有小村庄大小。

旧金山一夜之间变为一个城市是由于 1848 年在它附近发现了金子,致使大批人涌向加利福尼亚。马车缓缓驶过蔓延 2000 英里的草原和山川的危险地带,同时上百艘船绕过合恩岛在海上进行同等危险的航行。到加利福

47

尼亚时，成千上万的人下了船，船员们抛弃了他们的航船，上百艘船就这样被丢弃在海湾中直至腐烂。在两年内，加利弗尼亚就拥有了足以构成一个州的人口，而旧金山在几年内成为了这批外来人口的中心区。

这个城市的名声源于其良好的气候条件、安逸的生活、美食和众多的旅游景点。它以其丁当作响地爬上陡峭山崖的缆车及码头沿岸摆摊出售的海鲜食品而闻名。大多数来自太平洋沿岸的各国游客用几天的时间来了解这座城市。

47.【答案及解析】K）空格处应填入形容词用来修饰 Spanish outpost，由下句话中的 "The town was little more than a village" 可知旧金山最初和一个小村庄没有差别，对应到本句话中应该是说它是西班牙的一个小村落。

48.【答案及解析】L）空格处应填入名词，与 take...of 构成动词短语作谓语，take possession of 表示"占领，占有"。该句表明在与墨西哥的战争中，美国占领了旧金山，因此答案为 L）。

49.【答案及解析】B）空格处应填入动词作谓语，并可以与介词 into 搭配，句中 "into a city overnight" 是说旧金山从城镇一夜间发展成为城市，表明该动词应该是与突然的转变有关，选项中只有 sprang 符合题意。

50.【答案及解析】O）空格处应填入形容词用来修饰名词 rush，此句是说加利弗尼亚发生的变化，该形容词是用来修饰蜂拥而入的程度，"great"一词体现出来蜂拥而入的程度应该是非常大的，选项中 great 符合题意。

51.【答案及解析】M）空格处应填入形容词修饰 way，句中 "plodded" 说明行进过程的缓慢和艰难，即 "across 2,000 miles of prairie and mountains" 的过程应该不是一帆风顺的，而是带有危险性的，这可以从句中提到的 "hazardous trip" 看出路程的危险，选项中 dangerous 符合题意。

52.【答案及解析】I）空格处应填入副词用来修饰 hazardous，while 表明前后两句是对比关系，前半句是说乘坐马车是危险的，后半句是说乘坐帆船的旅程，句中 "around the Horn" 说明该过程也是危险的，要想表达与上句同样肯定的语气，选项中只有 equally 符合题意。而 hopefully 表示"充满希望地"，不符合题意。

53.【答案及解析】E）空格处应填入动词作谓语，由后半句 "hundreds of vessels were left to rot" 可知，这些船应该是被丢弃的，因此，选项中 deserted 符合题意。

54.【答案及解析】D）空格处应填入名词，加利福尼亚应该是有足够的人口构成一个州，后半句进一步说明旧金山多年来一直是人口集中地。因此，只有 state 符合题意。

55.【答案及解析】H）空格处应填入名词，due to 表明后面的不定式是导致该名词目前情况的原因，不定式是说"良好的气候、安逸的生活、美食和众多的旅游景点"，这些应该是形成城市繁华景象的原因，所以应选择 popularity，同时排除干扰项 "pollution 污染"。

56.【答案及解析】G）空格处应填入形容词用来修饰 hill，选项中能用于形容山脉的只有 steep "陡峭的"。

Section B

Passage one

Do you find getting up in the morning so difficult that it's painful? This might be called laziness, but Dr. Kleitman has a new explanation. He has proved that everyone has a daily energy cycle.

During the hours when you labour through your work you may say that you're "hot". That's true. The time of day when you feel most energetic is when your cycle of body temperature is at its peak. For some people the peak comes during the forenoon. For others it comes in the afternoon or evening. No one has discovered why this is so, but it leads to such familiar monologues as

57. If a person finds getting up early a problem, most probably _____.

A) he is very lazy

B) he refuses to follow his own energy cycle

C) he is not sure whether his energy is low

细节
推断
D) he is at his energy peak in the afternoon or evening

58. Which of the following may lead to family quarrels according to the passage?

同义
转述
A) Unawareness of energy cycles.

B) Familiar monologues.

48

"Get up, John! You'll be late for work again!" The possible explanation to the trouble is that John is at his temperature-and-energy peak in the evening. Much family quarrelling ends when husbands and wives realize what these energy cycles mean, and which cycle each member of the family has.

You can't change your energy cycle, but you can learn to make your life fit it better. Habit can help, Dr. Kleitman believes. Maybe you're sleepy in the evening but feel you must stay up late anyway. Counteract your cycle to some extent by habitually staying up later than you want to. If your energy is low in the morning, but you have an important job to do early in the day, rise before your usual hour. This won't change your cycle, but you'll get up steam and work better at your low point.

Get off to a slow start which saves your energy. Get up with a leisurely yawn and stretch. Sit on the edge of the bed a minute before putting your feet on the floor.

Avoid the troublesome search for clean clothes by laying them out the night before. Whenever possible, do routine work in the afternoon and save tasks requiring more energy or concentration for your sharper hours.

接58A

C) A change in a family member's energy cycle.

D) Attempts to control the energy cycle of other family members.

59. If one wants to work more efficiently at his low point in the morning, he should_____.

A) change his energy cycle

B) overcome his laziness

同义转述 C) get up earlier than usual

D) go to bed later

60. You are advised to rise with a yawn and stretch because it will_____.

同义转述 A) help to keep your energy for the day's work

B) help you to control your temper all the day

C) enable you to concentrate on your routine work

D) keep your energy cycle under control all day

61. Which of the following statements is NOT TRUE?

A) Getting off to work with a minimum effort helps save one's energy.

细节推断 B) Dr. Kletman explains why people reach their peaks at different hours of day.

C) Children have energy cycles, too.

D) Habit helps a person adapt to his own energy cycle.

【短文大意】你是否发现早晨起床是如此困难，以至于它都成了一件很痛苦的事了吗？这也许可以被称作懒惰，但是克莱曼博士有了一种新的解释。他已经证实每个人都有自己的能量循环。

在你努力工作期间，你也许会说你感觉很热。你的感觉是对的。一天中你感觉精力最充沛的时间正是你的体温循环处于高峰期的时刻。对某些人来讲，体温循环的高峰期出现在上午。而对其他人来说，它会在下午或晚上出现。没有人知道为什么会这样，但它却让我们常说如"约翰，起床了！你上班又快迟到了！"对这种问题可能的解释是约翰体温和能量循环的高峰出现在夜晚。许多家庭在吵架结束后，双方才意识到这些能量循环意味着什么，每个家庭成员拥有哪种能量循环。

你不能改变你的能量循环，但你可以学会让你的生活更好地适应它。克莱蔓博士认为习惯可以帮助你。也许你在晚上很困倦但无论如何你必须得熬夜。从某种程度上说养成晚睡一会儿的习惯可以适当调节你的能量循环。如果早晨你的能量循环处于低谷，但是你还有重要的工作要做，可以比平时早起床。这不会改变你的能量循环，但可以使你在能量循环低谷期时振作起精神并更好地工作。

缓慢地开始一天可以节省你的能量。起床时你可以慢慢地打个哈欠，伸伸懒腰。先在床边坐一分钟，然后再把脚放在地板上。避免在昨夜乱放的衣服中寻找干净的衣服而带来的麻烦。只要有可能，将日常工作留到下午去做，把那些需要更多精力或注意力的工作留到你处于能量循环高峰期时再去完成。

57. 【题目译文】如果一个人发现早起是个问题，最有可能是_____

【答案及解析】D) 第一段克莱曼博士对此的解释是每个人都有自己的能量循环。言下之意是，他早上不在状态，所以推断他下午或晚上会达到能量循环的顶点。故 D)为正确答案。

58.【题目译文】通过短文，下面哪个选项也许会导致家庭争吵？

【答案及解析】A) 见文中第二段最后一句话 "Much family quarrelling ends when husbands and wives realize what these energy cycles mean, and which cycle each member of the family has." 即：许多家庭在吵架结束后双方才意识到这些能量循环意味着什么，每个家庭成员有哪种循环。因此家庭吵架的原因是没有意识到这些能量循环意味着什么，而非一个家庭成员能量循环的改变或试图控制其他家庭成员的能量循环，故而排除 C)、D)。文中没有谈到 B)，因此 A)为正确答案。

59.【题目译文】如果一个人在早晨处于能量循环的低谷时想更有效地工作，他应该_____。

【答案及解析】C) 见原文第三段的倒数第二句话：你的能量循环在早晨处于低谷时，但是你还有重要的工作要做，你可以比平时早起床。这不会改变你的能量循环，但可以使你在能量循环低谷时有精力并更好地工作。故 C) 为正确答案。

60.【题目译文】建议在起床时打打哈欠和伸伸懒腰，因为它将_____。

【答案及解析】A) 见原文第四段第一、二句话 "Get off to a slow start which saves your energy. Get up with a leisurely yawn and stretch." 即你可以通过在起床时慢慢地打个哈欠，伸伸懒腰来开始慢慢地积累你的能量。因此 A)为正确答案。

61.【题目译文】下面陈述中哪一个是不正确的？

【答案及解析】B) 第一段克莱曼博士指出每个人都有自己的能量循环，当然包括孩子，即选项 C)正确。A)在原文第四段提到。第三段谈到习惯可以帮助你适应能量循环，即选项 D)正确。而原文根本没有提到为何每个人的能量顶点不一样，所以选项 B) 的意思与原文不符。

Passage Two

Reading to oneself is a modern activity that was almost unknown to the scholars of the classical and medieval worlds, while during the fifteenth century the term "reading" undoubtedly meant reading aloud. Only during nineteenth century did silent reading become commonplace.

One should be wary, however, of assuming that silent reading came about simply because reading aloud is distraction to others. Examination of factors related to the historical development of silent reading reveals that it became the usual mode of reading for most adult reading tasks mainly because the tasks themselves changed in character.

The last century saw a steady gradual increase in literacy, and thus in the number of readers. As readers increased, so the number of potential listeners declined, and thus there was some reduction in the need to read aloud. As reading for the benefit of listeners grew less common, so came the flourishing of reading as a private activity in such public places as libraries, railway

63. The development of silent reading during the nineteenth century indicated_____.

A) a change in the status of literate people

细节推断 B) a change in the nature of reading

C) an increase in the average age of readers

D) an increase in the number of books

62. Why was reading aloud common before the nineteenth century?

A) Reading aloud had been very popular.

B) There were few places available for private reading.

细节推断 C) Few people could read for themselves.

D) People relied on reading for enjoyment.

64. Educationalists are still arguing about_____.

A) the importance of silent reading

B) the amount of information obtained by books and newspapers

C) the effects of reading aloud and silent reading

carriages and offices, where reading aloud would cause distraction to other readers.

综合理解

Towards the end of the century there was still considerable argument over whether books should be used for information or treated respectfully, and over whether the reading of material such as newspapers was in some way mentally weakening. Indeed this argument remains with us still in education. However, whatever its virtues, the old shared literacy culture had gone and was replaced by printed mass media on the one hand and by books and periodicals for a specialized readership on the other.

By the end of the century students were being recommended to adopt attitudes to books and to use skills in reading them which were inappropriate, if not impossible, for the oral reader. The social, cultural, and technological changes in the century had greatly altered what the term "reading" implied.

D) The value of different types of reading material

65. The emergence of the mass media and of specialized periodicals showed that_____.

细节推断

A) standards of literacy had declined
B) reader's interests had diversified
C) printing techniques had improved
D) educationalist's attitudes had changed

66. What is the writer's purpose in writing this passage?

A) To explain how present-day reading habits developed. （主旨题）
B) To change people's attitudes to reading
C) To show how reading methods have improved
D) To encourage the growth of reading.

【短文大意】默读是一种就连遥远古和中世纪时期的学者都不知道的现代行为，而在 15 世纪，"阅读"这个术语无疑意味着大声朗读。默读仅仅在 19 世纪才逐渐普及。

然而，我们仍不能轻率地认为默读的产生是因为大声朗读会令他人分心。在对与默读的发展历史有关的种种因素的核查中显示，默读之所以成为大多数成年人的普遍阅读方式，是因为阅读任务本身的性质改变了。

上个世纪，随着读写能力的逐渐增强，阅读者的数量也在不断增加。因为阅读者的数量增加了，所以那些潜在的听者的数量就随之下降了，这样，对于大声朗读的需求也就减少了。随着使听者受益的阅读逐渐减少，在公共场所如图书馆、火车车厢和办公室，默读作为一种私人的活动就越来越兴盛了，因为在这些地方朗读会使其他读者分神。

一直到本世纪末，在书籍应当作为信息来源还是应该被尊敬地对待，阅读资料如报纸是否在某种程度上减弱精神作用方面仍存在着大量的争议。其实在我们的教育中也存在这种争论。然而，不论有怎样的优点，阅读能力共享的传统文化已经被各种印刷媒体和拥有专门读者的书刊所代替。

到本世纪末，学生们被建议采用不同的态度和运用各种阅读技巧去阅读书籍。因为有些书籍可能不适合于口头朗读。在本世纪，社会、文化、技术的变化极大地改变了阅读这个术语所包含的内涵。

62. 【题目译文】为什么 19 世纪前朗读十分普及？

【答案及解析】C) 根据文章的主题和第三段的内容，19 世纪人们的阅读能力不断提高，所以朗读的必要性逐渐减低。由此推断，在此之前朗读盛行的原因是人们阅读能力不高。故 C) 为正确答案。

63. 【题目译文】19 世纪默读的发展表明_____。

【答案及解析】B) 见文中第二段 "Examination of factors related to the historical development of silent reading reveals that it became the usual mode of reading for most adult reading tasks mainly because the tasks themselves changed in character." 即：在对与默读的发展历史有关的种种因素的核查中显示，默读之所以成为大多数成年人的普遍阅读方式，是因为阅读任务本身的性质改变了。因此 B) 为正确答案。

64. 【题目译文】教育家仍在争论_____。

【答案及解析】D) 文中第四段指出，对于书籍应当作为信息来源还是应该被尊敬地对待、阅读资料如报纸是否在某种程度上减弱精神作用方面仍存在着大量争议。实际上他们是在争论阅读材料（如书、报纸等）的价值，而且在我们的教育中仍然存在这种争论。故 D) 为正确答案。

65．【题目译文】大众传媒和专业杂志的出现表明_____。

【答案及解析】B) 由第四段最后一句可知，阅读能力共享的传统文化已经被各种印刷媒体和拥有专门读者的书刊所代替。这一点表明，读者的兴趣已经多样化，因此 B)为正确答案。

66．【题目译文】作者写这篇文章的目的是什么？

【答案及解析】主旨题。A) 本文意在分析阅读方式的历史演变及其原因，这也是文章的主题。B)改变人们对阅读的态度，C) 讲述阅读方法是如何改进提高的，D)鼓励阅读的发展均不符合文章的内容，因此 A)为正确选项。

Part V   Error Correction

【短文大意】人类一出现，就注定在这个星球上繁衍生息。现在人类已经具备了离开这个地球并且向宇宙拓展的能力了。而在以前，人类只能间接地了解宇宙。人类已经开发了月球的一部分，将宇宙飞船发射升空，让它沿着轨道绕另一个星球运转，10 年内将有可能登陆开发另一星球。我们能否大胆地设想一下在不远的将来我们能够征服其他的星球？一些人已经宣称这个计划可以解决人口问题：将多出来的人送上月球。但是我们必须知道我们将会花费数十亿开展这一项计划。为了使地球的人口维持在现在的水平，我们需要每天或者每小时向空间运送 7,500 人。

为什么我们要在空间拓展上花费这么多的钱呢？考虑到对改善地球各方面环境的极大要求，一个人对将金钱和资源源源不断地花费在空间拓展上的结果的关心是不无道理的。但是，或许我们在匆忙得出结论前应该考虑好事情的两面性。

67．时态。had→ has，根据全文时态可以看出此处应用一般现在时而非一般过去时，并且本句中有明显的时间副词 now。有些考生考虑到最后两句都用了现在完成时，因而在这里也误用了现在完成时 has had。其实，在表示"拥有"之意时，通常都只用动词的一般式，而不是其完成式。

68．文章的理解。directly →indirectly，根据该句意思，以前人们对宇宙的了解是间接的。

69．介词。into→on，本题所在句子的意思是：人类有可能在未来的 10 年中登上别的星球。表示"登上，在……上着陆"的意思时，动词 land 后面通常用介词 on。介词 into 的意思是"进入"，用在本句中显然欠妥。

70．短语搭配。too→so，本句的意思是：我们会不会如此地勇敢，以至于认为……。英语中表示"……到这种程度以至于……"时一般用短语 so...as to...，英语中没有 too...as to...这样的搭配，因而是错的。

71．指代一致。planet→ planets，代词 other 同时与复数名词连用，因而本句中的 planet 须用复数 planets。other 只有和定冠词连用时，才有可能用来表示单数，但通常是指两件事物中的另一件或两个人中的另一个。

72．固定搭配。head→mind，keep...in mind 是固定用法，表示"记住"。

73．上下文的理解。little→much，根据上下文，本句的意思应该是：我们为什么要在宇宙探索方面花这么多的钱？但这里用了 little 这个词，显然将意思弄反了。因此必须将句中的 little 改成 much，才符合上下文的逻辑。

74．句子结构。Consider→Considering，本题所在的句子中有两个并列的单句，却明显地没有连接词加以连接，这就需要考虑将其中的一个单句变成分词短语或介词短语，最合适的做法就是将动词 Consider 改成介词 Considering。介词 considering 的意思是"考虑到"。

75．句子结构。They→删去，本题所在的定语从句中主语与关系代词重复，二者必须去掉一个。从结构上来看，该去掉的只能是重复的主语 they。如果去掉关系代词 that，句子显然就不通了。

76．固定搭配。arriving→arriving at，arrive 是不及物动词。arrive at 是个固定搭配，表示"（经过努力）得出（结论），做出（决定）"。

Part VI Translation

77. It is simple. The more preparation you do now, _____(考试前你就越不会紧张).

【题目译文】这很容易，你现在准备得越充分，考试前你就越不会紧张。

【答案及解析】the less nervous you'll be before the exam 注意比较级 the more …, the more …的用法。

78. She often complains about not _____(感到自己在工作上不受赏识).

【题目译文】她常常抱怨，感到自己在工作上不受赏识。

【答案及解析】feeling appreciated at work 短语 "complain doing sth." 抱怨做某事。

79. The results of the exam will be _____(推迟到) Friday afternoon.

【题目译文】考试的结果将推迟到周五的下午。

【答案及解析】put off to 注意短语的用法。

80. It turns out that the price _____（开始回落）.

【题目译文】结果是价格开始回落。

【答案及解析】begins to go down 注意短语 "go down" 的用法。

81. _____(据说) a foreign teacher will come to our class.

【题目译文】据说一位外教老师将要来到我们的班级。

【答案及解析】It is said that 注意句式 "It is said that…": 据说……

参考答案

Part I Writing

Wealth and Happiness

Wealth has always been what some people long for. It is true that most of them try to get wealth by means of honest labor. Their efforts contribute to the welfare of the society and at the same time to the accumulation of their wealth, and hence their happiness.

There is no doubt that wealth brings happiness especially in the modern society, where various kinds of modern conveniences, new fashions and entertainments make their appearances with each passing day to make life more comfortable and colorful. Only money can turn admiration into reality.

But there are exceptions when wealth does not go hand in hand with happiness. Wealth may encourage those weak-willed persons to be addicted to some harmful habits, such as drug-taking or gambling, and bring about their own ruin. Also, a person may lose his reason and go astray if he is passionately devoted to seeking wealth. Therefore, one can never reckon only on wealth to achieve happiness.

Part II Reading Comprehension (Skimming and Scanning)

1. Y   2. N   3. NG   4. N   5. Y   6. N   7. Y

8. complaint

9. Education Inspection and Review Request Form

10. height, weight, sport of participation

Part III Listening Comprehension

Section A

11. C   12. A   13. D   14. B   15. A   16. B   17. D   18. D

19. C   20. A   21. C   22. A   23. A   24. B   25. A

Section B

26. C 27. B 28. B 29. D 30. C 31. A 32. A 33. D 34. B 35. C

Section C

36. refers   37. appeals   38. metaphor   39. negatively

40. expression 41. consumption 42. especially   43. derived

44. Later the increasing employment of women in offices led to the introduction of afternoon tea, in which gradually the male members of a staff would join.

45. Individual tastes, however, often varied: for China or India tea; weak or strong; with or without milk; with or without sugar; with or without lemon.

46. "That is not my cup of tea", and then by extension in general reference to other things that did not suit one's taste

Part IV Reading Comprehension (Reading in Depth)

Section A

47. M: reporters   48. I: basic   49. C: know   50. A: inventions   51. G: improve

52. E: fields   53. H: serious   54. K: economic   55. D: heavily   56. O: small

Section B

57. C   58. A   59. D   60. A   61. B   62. C   63. C   64. B   65. C   66. D

Part V   Cloze

67. C   68. A   69. B   70. A   71. A   72. B   73. B   74. D   75. C   76. C

77. D   78. A   79. B   80. A   81. C   82. A   83. B   84. C   85. B   86. B

Part VI   Translation

87. moreover   88. have taken a lot of courage

89. at her best   90. admit they are wrong   91. rather than a teacher

答案解析及录音原文

Part I   Writing

Wealth and Happiness

Wealth has always been what some people long for. It is true that most of them try to get wealth by means of honest labor. Their efforts contribute to the welfare of the society and at the same time to the accumulation of their wealth, and hence their happiness.

There is no doubt that wealth brings happiness especially in the modern society, where various kinds of modern conveniences, new fashions and entertainments make their appearances with each passing day to make life more comfortable and colorful. Only money can turn admiration into reality.

But there are exceptions when wealth does not go hand in hand with happiness. Wealth may encourage those weak-willed persons to be addicted to some harmful habits, such as drug-taking or gambling, and bring about their own ruin. Also, a person may lose his reason and go astray if he is passionately devoted to seeking wealth. Therefore, one can never reckon only on wealth to achieve happiness.

【评论】这篇文章主要是谈论财富和幸福的关系问题。这样的文章通常分为三段来写。在本文中，第一段谈到有些人努力奋斗获得财富，第二段谈到财富有时的确能给人带来幸福，第三段从反面去议论有时光有财富并不能给人带来幸福。这篇文章从正反两方面去论述，条理清楚，令人信服。

从正反两方面谈论问题时常用到如下句式及关联词：

第一段：点明主题

It is true that most of them try to do..., their efforts contribute to..., at the same time, hence.

第二段：从正面论述

There is no doubt that...

第三段：从反面论述

But there are exceptions when..., such as..., also, therefore

Part II   Reading comprehension (Skimming and Scanning)

【文章及答案解析】本文是对大学如何执行"Policy on Student Privacy Rights"(学生隐私权政策声明)的阐述，主要向大学生介绍了大学生教育记录和学生登录信息的相关规定。

Policy on Student Privacy Rights Policy Statement

Under the Family Educational Rights and Privacy Act (FERPA), you have the right to:

● inspect and review your education records;

● request an amendment to your education records if you believe they are inaccurate or misleading;

● request a hearing if your request for an amendment is not resolved to your satisfaction;

● consent to disclosure of personally identifiable information from your education records, except to the extent that FERPA authorizes disclosure without your consent;

● file a complaint with the U.S. Department of Education Family Policy Compliance Office if you believe your rights under FERPA have been violated.

1. Inspection

What are education records?

Education records are records maintained by the university that are directly related to student. These include biographic and demographic data, application materials, course schedules, grades and work-study records. The term does not include:

● information contained in the private files of instructors and administrators, used only as a personal memory aid not accessible or revealed to any other person except a temporary substitute for the maker of the record;

● Campus Police records;

● employment records other than work-study records;

● medical and psychological records used solely for treatment purposes;

● records that only contain information about individuals after they have left the university;

● any other records that do not meet the above definition of education records.

【文章大意】　　学生隐私权政策

政策声明

根据家庭教育权和隐私权法案(FERPA)，你有权：

●检查和审查你的教育记录；

●认为自己的教育记录出现不实或存在误解的地方，可以申请修正，你的教育记录；

主旨大意

1. This article has university students as its target audience. (Y)

细节辨认

2. Under FERPA, students are entitled to request an amendment to their education records whenever they wish. (N)

●在所提出的要求没有得到满意解决的情况下,可以要求召开听证会；

●同意公布教育记录中关于个人身份的资料，除不需本人同意而由 FERPA 授权公布的范畴以外；

●如果认为自己的权利在 FERPA 中受到侵犯,可以向美国联邦教育部家庭政策执行处进行投诉。

1. 检查

什么是教育记录？

细节辨认

3. The education records are kept by the students themselves. (NG)

大学保存的教育记录是直接与学生有关的记录，包括个人及人口的统计数据、申请材料、课程安排、成绩以及工作学习记录。该记录不包括：

●仅用来作为私人辅助记忆而保存在管理员和教员私人档案中的，除临时代替的记录者以外，不可以向其他任何人透露的信息；

●校园警察的记录；

●不同于工作学习记录的就业记录；

●仅以医治为目的的医疗和心理治疗记录；

●离校后，仅存有个人信息的记录；

●不符合上述教育记录定义的其他记录。

How do I inspect my education records?

● Complete an Education Inspection and Review Request Form (available online as a PDF document or for The HUB, 12C Warner Hall) and return it to The HUB.

● The custodian of the education record you wish to inspect will contact you to arrange a mutually convenient time for inspection, not more than 45 days after you request. The custodian or designee will be present during your inspection.

● You will not be permitted to review financial information, including your parents' financial information; or confidential letters of recommendation, if you have waived your right to inspect such letters.

● You can get copies of your education records from the office where they are kept for 25 cents per page, prepaid

2. Amendment

How do I amend my educational records?

● Send a written, signed request for amendment to the Vice President for Enrollment, Carnegie Mellon University, 610 Warner Hall, Pittsburgh, PA 15213. Your request should specify the record you want to have amended and the reason for amendment.

● The university will reply to you no later than 45 days after you request. If the university does not agree to amend the record, you have a right to have a hearing on the issue.

3. hearing

How do I request a hearing?

Send a written, signed request for a hearing to the Vice President for Enrollment, Carnegie Mellon University, 610 Warner Hall, Pittsburgh, PA 15213. The university will schedule a hearing no later than 45 days after your request.

如何检查自己的教育记录？

9. Student who wants to inspect his education records must fill out an _____.
(**Education Inspection and Review Request Form**)

●填写教育查看回顾申请表（可从网上获得 PDF 文件或到位于华纳大楼 12C 的中心领取）并交回中心；

4. A student's demand for the inspection of his records must be met within a month. (**N**)

●提出申请的 45 天内，该教育记录的管理者会与你联系，安排彼此方便的时间进行检查，届时管理者或指定人将出席；

●如果你已被免除检查这类信函的权利，你将不被允许审查包括父母的财务状况或推荐密函在内的财政资料；

●你可以在办公室对你的教育记录进行影印，每页需支付 25 美分。

2．修改

如何修改我的教育记录？

●将要求修改的书面申请递交给管理注册的副校长，卡内基梅隆大学，华纳大楼 610、匹兹堡，PA15213。申请应详细说明想要修改的内容和修改的理由；

5．When the request for an amendment is refused by the university, the student may ask for a hearing. (**Y**)

●校方将在提出申请后的 45 天内给予答复。如果大学不同意修改记录，你有权对此问题召开听证会。

3．听证

如何申请听证？

●将书面申请递交给管理注册的副校长，卡内基梅隆大学，华纳大楼 610，匹兹堡，PA15213。大学将在提交申请后的 45 天内安排听证会；

How will the hearing be conducted?

● A university officer appointed by the President for Enrollment, who is not affiliated with your enrolled college, will conduct the hearing.

● You can bring others, including an attorney, to the hearing to assist or represent you. If your attorney will be present, you must notify the university ten days in advance of the hearing so that the university can arrange to have an attorney present too, if desired.

● The university will inform you of its decision, in writing, including a summary of the evidence presented and the reasons for its decision, no later than 45 days after the hearing.

●If the university decides not to amend the record, you have a right to add a statement to the record that explains your side of the story.

4. Disclosure

Carnegie Mellon generally will not disclose personally identifiable information from your education records without your consent except for directory information and other exceptions specified by law.

What is directory information?

Directory information is personally identifiable information of a general nature that may be disclosed without your consent, unless you specifically request the university not to do so. It is used for purpose like compiling campus directories.

If you do not want your directory information to be disclosed, you must notify The HUB, 12C Warner Hall, in writing within the first 15 days of the semester.

Notifying The HUB covers only the disclosure of centralized records. Members of individual organizations such as fraternities, sororities, athletics, etc. must also

如何进行听证?

●由副校长指定的不属于你所在注册学校的大学官员主持听证;

●你可以带包括一名律师在内的其他人帮助或代表你出席听证会。如果有律师出席,你必须提前 10 天通知大学,以便大学在必要的情况下也安排一名律师;

> 7. In a hearing for the amendment of education records, both the student and the university may hire attorneys. (Y)

同义转述

●听证会后的 45 天内,校方将以书面形式告知其决定,其中包括证词陈述的概要和做出此决定的缘由;

●如果大学决定不修改记录,你有权添加解释你方观点的记录。

4. 公开

除了联络资料和其他法律规定的信息以外,卡内基梅隆大学一般不会未经本人许可擅自泄露教育记录中的个人资料。

什么是联络资料?

联络资料是不经本人同意就可以公布于众的个人身份信息,除非你明确要求学校不予公开,它用于类似于校园通讯录编制这样的目的。

如果你不想你的联络资料被公开,你必须在本学期开始的 15 天内以书面形式通知位于华纳大楼 12C 的网络中心。

> 6. The university is free to disclose the directory information of students without their consent. (N)

细节推理

通知网络中心只公开主要的记录。私人组织机构如兄弟会、妇女联合会、田径会等的会员也必须通知其组织禁止公开联络资料。

notify those organizations to restrict the disclosure of directory information.

Carnegie Mellon has defined directory information as the following:

- your full name
- local/campus address
- local/campus telephone number
- email user id and address
- major, department, college
- class status (freshman, sophomore, junior, senior, undergraduate, or graduate)
- dates of attendance (semester begin and end dates)
- enrollment status (full, half, of part time)
- date(s) of graduation
- degrees awarded
- sorority or fraternity affiliation

For students participating in intercollegiate athletics, directory information also includes:

- height, weight
- sport of participation

What are the other exceptions?

Under FERPA, Carnegie Mellon may release personally identifiable information from your education records without your prior consent to:

- school officials with legitimate educational interests;
- certain federal officials in connection with federal program requirements
- organizations involved in awarding financial aid;
- state and local officials who are legally entitled to the information;
- testing agencies such as the Educational Testing Service, for the purpose of developing, validating, researching and administering tests;
- accrediting agencies, in connection with their

卡内基梅隆大学所规定的联络资料如下:

- 姓名全称
- 地方/学校住址
- 本地/校园电话号码
- 电子邮件的登录名和地址
- 专业、系、学校
- 班级情况（大一、大二、大三、大四、本科或研究生）
- 入学日期（学期开始日期和结束日期）
- 注册情况（全年，半年，业余）
- 毕业日期
- 学位授予
- 兄弟姐妹关系

对于参加校际田径比赛的学生，资料还包括:

- 身高、体重
- 参加的体育运动

10. The directory information for intercollegiate athletes include additional information such as_____. (**height, weight, sport of participation**)

什么是其他的例外?

根据 FERPA、卡内基梅隆大学还可以在未经许可之前公布教育记录中的个人身份信息给:

- 拥有合法权益的学校教育官员；
- 某些与国家需求计划有关的联邦官员；
- 包括在财政上给予支持的组织；
- 依法授权的州立和当地的官员；
- 以发展、验证、研究和管理测试为目的的检测机构如教育考试中心；
- 与评审有关的审批机构；
- 依赖父母的学生（美国国家税收局法规第 152 条对此做出了定义）；
- 健康或安全紧急事件中的相关团体，必要情况下保护学生或他人的健康或安全；

accrediting functions;

- parents of dependent students (as defined in section 152 of the Internal Revenue Service Code);
- appropriate parties in a health or safety emergency, if necessary to protect the health or safety of the student or other individuals;
- officials of another school in which the student seeks or intends to enroll;
- victims of violent crimes or non-forcible sexual offenses (the results of final student disciplinary proceedings);
- parents or legal guardians of student under 21 years of age (information regarding violations of university drug and alcohol polices);
- Courts (records relevant to legal actions initiated by students, parents or the university).

- 尝试或想注册的其他学校的官员;
- 暴力犯罪或非暴力性犯罪的受害者(学生纪律处分的终审结果);
- 未满 21 周岁学生的父母或法定监护人（关于违反大学生毒品和酒精条例的资料）;
- 法院（关于学生、家长或大学发起的合法行动的记录）。

8. If a student feels that his right under FERPA has been violated, he could file a _____ with the U.S. Department of Education Family Policy Compliance Office. (**complaint**)

5. **Complaints**

If you believe the university has not complied with FERPA, you can file a complaint with the:
Family Policy Compliance Office
Department of Education
400 Maryland Avenue, S.W.
Washington, DC20202-4605

5. 投诉

如果你认为大学没有遵守 FERPA，你可以进行投诉:
家庭政策执行处
教育部
马里兰大街 400 号
华盛顿哥伦比亚特区西南部 20202-460

1.【题意】本文以大学生为读者对象。

【解析】本题属于主旨题。本句的关键词是 university students。根据标题可知,本文是关于大学生如何执行"Policy on Student Privacy Rights"（学生隐私权政策）的说明文,对象是就读大学生。所以一看文章标题就可做出肯定的判断,故应选 YES。

2.【题意】根据 FERPA 法案,学生任何时候都可以要求修改他们的教育记录。

【解析】本句的关键词是 FERPA、entitle 及 whenever。entitled to(have the right to)告诉我们应去有关学生权利的部分进行定位。在"Policy Statement"部分,文章明确指出了要求修改教育记录的条件是:"if you believe they are inaccurate or misleading"而不是任何时候。本句不符合文章的意思,故选 NO。

3.【题意】教育记录由学生自己保管。

【解析】此句的关键词是 education records。可在文章中的"What are education records?"部分查找,一下就能找到"Education records"的有关信息。但原文并没有说由谁保管,所以判断为 NOT GIVEN。

4.【题意】学生要求查看记录的请求必须在一个月内得到答复。

【解析】此句的关键词是 within a month,据此去查找有关时间的表述。本文有三处出现了"45 days",按照从前往后的顺序进行 scan，首先在 "How do I inspect my education records?" 部分找到关于安排学生查看记录的时间期限的规定："The custodian of the education record you wish…, not more than 45 days after your request."，可见安排查看的时间是提出请求后的 45 天内，而不是 30 天，故本题应选的是 NO。

5．【题意】如果要求修改教育记录的要求遭到拒绝，学生有权申请听证。

【解析】此句的关键词是 amend，先去 "2. Amendment" 部分进行定位查找。在 "How do I amend my educational records？" 部分的最后一句找到了要求听证的条件："If the university does not agree to amend the record, you have a right to a hearing on the issue."，与原文意思一致，故本题应选的是 YES。

6．【题意】大学无需征得学生的同意即可随意公布他们的联系信息。

【解析】本句的关键词是 directory information，所以必须跳过 "How will the hearing be conducted？" 部分，去 "What is directory information？" 部分查找。 原文中指出学生可以要求不公开自己的联络信息："If you do not want your directory information to be disclosed, you must notify…"，可见本句与原文不符，故本题应选的是 NO。

7．【题意】在要求修改教育记录的听证会上，校方和学生均可聘用律师。

【解析】此句的关键词是 hearing 与 attorney 可到 "How will the hearing be conducted？" 部分去查找。在该部分的第二段提到 "如果你请了律师，应在 10 天前通知校方，好让校方也请律师"，可见双方均可请律师。本句符合文章的原意，故本题应选 YES。

8．【题意】如果一个学生觉得由 "家庭教育权利和隐私权法案" 所保护的权利受到侵害，可以向美国教育部家庭政策执行处进行投诉。

【解析】本句的关键词是 right、violated 与 file。答案在文章的开头与结尾两次出现，很明显这里要填 complaint。

9．【题意】需要查看自己教育记录的学生必须填写 "查看教育记录申请表"。

【解析】本句的关键词是 inspect。根据它的提示，应该去 "How do I inspect my education records?" 部分查找，由原文可知应填：Education Inspection and Review Request Form。

10．【题意】校际运动员的登录信息还包括身高、体重和参与的运动。

【解析】此句的关键词是 directory information 与 intercollegiate。由此可知应该去 "What is directory information？" 部分查找本题的出处。在该部分最后找到了 intercollegiate 一词，即找到了答案。

Part Ⅲ　　Listening Comprehension
Section A

11．W: Wasn't Carl supposed to give a speech tonight?

M: Yes, but he backed out at the last minute.

Q: What do we learn about Carl?

【答案与解析】动词短语意义辨析题。女士认为卡尔今晚要发表讲话，男士说他最后决定不做了。关键短语是 back out（收回，停止不干），说明答案是 C。

12．M: I thought it would be fun if we all went to see that new movie downtown.

W: Count me out. I've heard it's not worth the money.

Q: What does the woman mean?

【答案与解析】动词短语意义辨析题。男士建议大家一起去市区看电影，女士认为电影不值那么多钱，让他不要把自己算在内。关键短语是 count me out（别把我算在内），说明答案是 A。

13. W: I can't wait to see the look on Betty's face when she opens our gift.

M: Neither can I.

Q: What does the conversation tell us?

【答案与解析】判断推理题。女士急于见到贝蒂收到礼物时的表情，男士说他也急于见到。关键是男士的回答 Neither can I.（我也等不急了），说明答案是 D。

14. W: Jane has written some articles for her research project.

M: So she has finished them.

Q: What had the man assumed about Jane?

【答案与解析】判断推理题。女士说简已经为她的研究项目写了一些文章了，男士说看来她已经写完了，意味着他原来认为简并没写完，说明答案是 B。

15. M: I wonder if a problem like this can be solved by Linda.

W: Well. If she can't solve it, no one can.

Q: What can be concluded from the conversation?

【答案与解析】判断推理题。男士不知道琳达能否解决这样的问题，女士回答道如果她不能解决，可就没人能解决了，说明琳达的能力比别人强。关键是 no one can（无人能解决了），因此答案是 A。

16. W: I'm going to ask the janitor to clean the lobby.

M: Would you ask him to wash the staircase as well?

Q: What is being discussed?

【答案与解析】判断推理题。女士准备请门卫把走廊打扫一下，男士说能否让他也清扫一下楼梯，他们显然是在谈论大楼的卫生问题，说明答案是 B。

17. W: One piece of cloth is pure wool and the other is a synthetic fabric.

M: Amazing! I really can't tell them apart.

Q: What does the man mean?

【答案与解析】动词短语意义辨析题。女士指出两块面料一块是纯羊毛另一块是合成纤维。男士很惊讶，说他根本看不出有什么区别。关键是男士的话语 I really can't tell them apart.（我真的无法把它们区分开来），说明答案是 D。

18. M: What was it you wanted to talk to me about just now?

W: Never mind, it was nothing important.

Q: What will the man probably do?

【答案与解析】判断推理题。男士问女士刚才想跟他谈什么，女士说没关系，不重要，显然男士刚才没听到女士的话，他应该对此向女士道歉，所以答案是 D。

Now you will hear two long conversations.

Conversation One

W: Jim, thank goodness, you've arrived ! The class presentation started half an hour ago. And I was just beginning to panic!

M: Sorry to be late, Ellen. This morning has been a real mess. I didn't think I was able to make it here at all.

W: Why are you late? Our whole presentation depends on the graphs you're holding.

M: Yes, I know. I'll tell you about it later. Two groups are still ahead of us, I think. Their presentation on the rights of the consumer and analysis of the stock market.

W: You do look cold! What happened?

M: I've been standing outside in arctic temperature for over an hour, waiting for the bus.

W: Over an hour? But I thought your apartment was only a ten-minute bus ride to campus.

M: Under normal conditions. But the bus was delayed because of the weather. And when I stepped into a drug-store to call home for a ride, the bus went by. As luck would have it, there was no one at home, so I had to wait another 45 minutes for the next bus.

W: That's Murphy's Law, isn't it? If anything can go wrong, it will. Well, we've still got twenty minutes.

M: We'd better stop talking. People are turning around and looking at us.

Questions 19 to 22 are based on the conversations you have just heard.

19. What were the students doing when the man arrived in class?

【解析】事实细节题。The class presentation started half an hour ago. 文中提到班级演讲已开始半小时了，所以学生们正在做演讲。

20. What class were the man and woman probably taking?

【解析】事实细节题。根据…rights of the consumer and analysis of the stock market. 消费者权益和股市分析可知所选修的课程是商务类课程。

21. How did the man come here?

【解析】事实细节题。文中提到男士家离学校很近，通常只需 10 分钟的车程，但是今天的特殊情况使他在路上耽搁了一个多小时，而且不得不等了 45 分钟的公交车。

22. What is the Murphy's law?

【解析】事实细节题。莫非法则是一种幽默的规则：任何可能出错的事终将出错。

Conversation Two

M: Are you ready for the trip to the museum in the Big Apple? I can hardly wait.

W: The Big Apple? What are you talking about?

M: The Big Apple is the nickname for New York City. You are going to New York with us, aren't you?

W: Yes, I am, especially looking forward to seeing the Museum of Modern Art. There is a special show of 20th century American Painters there. But tell me, where did the nickname come from?

M: The jazz musicians of the 1920s are responsible for the name. When they played a concert in a city, they called that city an apple. Of course, New York was the biggest city in the country and the best place for a jazz concert. So, the musicians called it the Big Apple.

W: Amazing! New York is such a fascinating place and it even has an interesting nickname, one that it has had for more than 50 years.

Questions 23 to 25 are based on the conversations you have just heard.

23. What was the woman interested in seeing?

【解析】事实细节题。根据已给选项的信息，判断本题四个选项都是关于文化的名词，与女士所说的现代艺术馆最贴近的就是 A。

63

24. Who gave New York its nickname?

【解析】事实细节题。根据已给选项的信息，可推断出本题是关于人物或职业的题目，男士话中的 are responsible for 给了本题的答案提示，jazz musicians 与 B 项同义。

25. How did the woman describe New York?

【解析】事实细节题。根据对最后一句话的了解，New York is such a fascinating place... 可推知选 A。

Section B

Passage One

We do not know when men first began to use salt, but we do know that it has been used in many different ways throughout history. Historical evidence shows, for example, that people who lived over 3,000 years ago ate salted fish. Thousands of years ago in Egypt, salt was used to keep the dead from decaying.

Stealing salt was considered a crime during some periods of history. In the 18th century, for instance, if a person were caught stealing salt, he could be put in jail. History records that about ten thousand people were put in jail during that century for stealing salt! About 159 years before, in the year 1553, taking more than one's share of salt was punished as a crime. The offender's ear was cut off!

Salt was an important item on the table of a king. It was traditionally placed in front of the king when he sat down to eat. Important guests at the king's table were seated near the salt. Less important guests were given seats farther away from it.

Questions 26 to 29 are based on the passage you have just heard.

【内容概要】本段听力材料讲的是人类使用食盐的历史以及许多相关的趣闻。三千多年以前人们就开始食用腌制的咸鱼了，偷盗食盐在历史上的许多时期都被视为一种犯罪，盐也是皇帝餐桌上的重要物品。

26. How was salt used in Egypt thousands of years ago?

【解析】细节考查题。根据第一段最后一句话 "Thousands of years ago in Egypt, salt was used to keep the dead from decaying"（几千年前在埃及，盐就被用来防止尸体腐烂）得知答案是 C。

27. Why were the ten thousand people put in jail in the 18th century?

【解析】细节考查题。根据第二段第二句话 "In the 18th century, for instance, if a person were caught stealing salt, he could be put in jail"（例如，在 18 世纪，如果一个人偷盐被抓，他会被送进监狱）得知答案是 B。

28. In the 16th century what would happen to a man who took more than his share of salt?

【解析】细节考查题。根据第二段最后两句话 "About 159 years before, in the year 1553, taking more than one's share of salt was punished as a crime. The offender's ear was cut off"（159 年以前，也就是 1553 年，多拿一份盐就会被作为罪犯受到惩罚。犯罪人的耳朵会被割下来）得知答案是 B。

29. When did man first begin to use salt?

【解析】细节考查题。根据文章第一句话 We do not know when men first began to use salt... （我们不知道人类从什么时候开始使用盐……）可知答案是 D。

Passage Two

Women's fashions tend to change more rapidly and radically than men's. In the early 1900s, all women wore their skirts down to the ankle. Today, skirt length varies from floor-length to ten inches above the knee. Women's shoes have also gone through all sorts of changes in the last ninety years. For example, boots for women were very common around the turn of the 20th century. Then, for years, they were not considered fashionable. Today they are back in style again in all colors, lengths, and materials. In fact, today's women can wear all sorts of clothes, even slacks and shorts, on almost any occasion. While all of these changes were taking place in women's fashions, men's clothing remained pretty much the same until a couple of years ago. In fact, most men still wear the traditional suit though bright colors and varieties in cut are now more common.

Questions 30 to 32 are based on the passage you have just heard.

【内容概要】本段听力材料讲的是与男士服装相比，女士服装的变化更为迅速和激进。20世纪初，女士的裙子都长及脚踝，现在则长短不一。与之大相径庭的是，男士的服装没有什么大的变化，大多数男士仍以传统的西装为主。

30．How did the women in the early 1900s wear their skirts?

【解析】细节考查题。根据第二句话 "In the early 1900s, all women wore their skirts down to the ankle"（在20世纪早期，女士的裙子都长及脚踝）得知答案是C。

31．What kind of shoes was considered fashionable for women at the turn of the 20th century?

【解析】细节考查题。根据段中的 "For example, boots for women were very common around the turn of the 20th century"（例如，女士的皮靴在20世纪交接时非常普遍）得知答案是A。

32．Which of the following best express the main idea of the talk?

【解析】综合分析题。根据全文谈论的内容，本段材料主要讲述的是随着时间的变化，女士与男士服装变化的不同，因此选择答案A。

Passage Three

Once again in Richfield Heights today, there were no classes in the public elementary and secondary schools. This marks the eleventh day that the schools in that community have remained closed. Teachers are still on strike despite the back-to-work order issued yesterday by the District Court. A spokesperson for the Teacher's Union stated at a press conference today that the strike would continue until such time as the School Committee agrees to a public hearing to settle the dispute. According to the spokesperson, wages are not an issue in the dispute. An issue is a new rule passed by the School Committee, which eliminates paid sick leave from the teachers' contract. Under the new contract, teachers would receive no salary for any day on which they failed to appear due to illness. School administrators, on the other hand, would continue to receive fifteen days of paid sick leave annually. The Parent Board, which was initially sympathetic to the teachers' position, has urged the teachers to return to work until a settlement can be reached. Spokespersons for the School Committee are refusing to comment on the latest developments.

Questions 33 to 35 are based on the passage you have just heard.

【内容概要】本段听力材料讲的是某地（Richfield Heights）因教师罢课导致中小学停课的事情。此次争端的主要问题不是工资问题，而是学校委员会在教师合同中取消了教师带薪病休的规定。

33．What seems to be the main issue in the dispute?

【解析】细节考查题。文中说明争端的主要问题不是工资问题，而是学校委员会在老师合同中取消了教师带薪病休（paid sick leave）的规定，所以答案是D。

34．What is the present position of the Parent Board regarding the strike?

【解析】细节考查题。根据文中的 "The Parent Board, which was initially sympathetic to the teachers' position, has urged the teachers to return to work until a settlement can be reached"（家长委员会开始很同情老师的遭遇，但现在

他们希望争端解决后，老师马上回到工作岗位）判断得知答案是 B。

35. According to the news, which of the following institutions is LEAST likely to be in favor of the back-to-work order?

【解析】综合分析题。根据全文的内容，再加上文中明确指出的 "A spokesperson for the Teacher's Union stated at a press conference today that the strike would continue until such time as the School Committee agrees to a public hearing to settle the dispute."（教师联盟的发言人在新闻发布会上声称罢工会继续，直到学校委员会同意开听证会，解决此争端为止。）可见 the Teacher's Union 最不可能支持回学校工作的命令，由此可判断答案是 C。

Section C

My cup of tea (36) refers to the sort of thing that pleases or (37) appeals to me. The (38)metaphor is nearly always used (39)negatively. The (40)expression came into use between the First and Second World Wars. In the Victorian age the (41)consumption of tea by all classes had not yet, (42)especially among men, become common. A more likely metaphor then, (43)derived from food or drink for something not to one's taste, would have been, say, "not my pot of beer", or among the well-to-do classes, "not my glass of wine".

(44)Later the increasing employment of women in offices led to the introduction of afternoon tea, in which gradually the male members of a staff would join. Later, tea would come to be regarded as a universal social drink. (45)Individual tastes, however, often varied: for China or India tea; weak or strong; with or without milk; with or without sugar; with or without lemon. This variation would naturally lend itself to the expression: " (46)That is not my cup of tea", and then by extension in general reference to other things that did not suit one's taste: e.g. an entertainment at a theatre, a book, etc, with the meaning "Whatever others may like, that is not the sort of thing to appeal to me".

Part IV   Reading Comprehension (Reading in Depth)
Section A

【短文大意】当某一事件发生后不久，遍布街头的报纸就会传递给大众详尽的信息。无论事件发生在世界的哪个角落，记者都会云集到事发现场去收集新闻。报纸有一个基本宗旨，就是迅速从事发现场和相关的人员那里获得新闻，报道给想要了解此事件的大众。

收音机、电报、电视和其他的发明都在和报纸进行竞争。杂志和其他传播媒介的发展也同样在和报纸竞争。然而竞争仅仅推动了报纸的发展。他们很快利用更新更快的信息传播方式来提高速度，因此他们的运营效率同样也提高了。今天，报纸的发行量和阅读的人数比以前多得多。竞争同时促进了报纸业向其他的领域拓展。今天的报纸除了让读者了解最新的新闻以外，它还在政治和其他重要事情上引导和影响读者们。报纸还通过广告影响着读者的经济选择。大部分报纸都是非常依赖商业广告得以生存的。

销售报纸所获得的钱不能支付全部甚至一小部分出版的费用，大部分报纸的收入主要来源于商业广告，广告做得是否成功取决于报纸对广告商的价值，这是用发行量来衡量的。

47.【答案及解析】M) 文章首句点出了文章的话题：newspapers。由空格后的系动词 are 可知所填词为复数名词作主语。与 newspaper 这个主题有联系的自然是 reporters，只有记者才会去收集新闻。

48.【答案及解析】I) 所填词修饰 purpose，为形容词，只有 basic 可以修饰 purpose，表示"基本宗旨"。

49.【答案及解析】C) 报纸旨在传播新闻，与 who make it 对应的是 who want to know it，即从新闻来源到新闻读者，而 produce 与 make 重复，体现不了消息的传播。

50.【答案及解析】A) 空格前的 radio、telephone 以及 television 都是一种发明，and other 说明了这种并列关系，因此选择 inventions。

51.【答案及解析】G) 所填词与 to 构成不定式，speed 为动词宾语。使用更快的信息传播方式，目的是提高速度，故选择 improve，排除 slow。

52.【答案及解析】E) 所填词由 other 修饰，为可数名词的复数。根据下文，当今报纸会教育读者，也会通过广告影响读者的决策，这些都是新闻报纸除了传播消息之外的其他作用，故选择 fields，表示报纸业还向其他领域拓展。

53.【答案及解析】H) 所填词与 important 一起修饰 matters，根据 and 表示的并列意义，选择与 important 并列的 serious。

54.【答案及解析】K) 所填词后为 choices，由此判断应填入形容词。根据后面 advertising，判断由广告影响的应该是经济决策，所以选择 economic。

55.【答案及解析】D) 空格中需要填入副词修饰 depend，将选择范围缩小到 heavily 和 slightly 上。而最后一段指出多数报纸的主要收入来源（main source of income）是商业广告，可见报纸的生存极为依赖广告。

56.【答案及解析】O) 所填词修饰 fraction，与 fail 和 even 一样都表示消极含义。选项中 small 修饰 fraction，并与后文的 main 形成对照。

Section B

Passage one

BMW's efforts to harness the creativity of its customers began two years ago when it posted a toolkit on its website. These toolkits let BMW's customers develop ideas showing how the firm could take advantage of advances in technology and in-car online services. From the 1,000 customers who used the toolkit, BMW chose 15 and invited them to meet its engineers inMunich. Some of their ideas (which remain under wraps for now) have since reached the prototype stage, says BMW. "They were so happy to be invited by us, and that our technical experts were interested in their ideas", says Mr Reimann. "They didn't want any money."

Westwood Studios, a game developer now owned by EA, first noticed its customers innovating its products after the launch of a game "Red Alert" in 1996: gamers were making new content for existing games and posting it freely on fan websites. Westwood made a conscious decision to embrace this phenomenon. Soon it was shipping basic game-development tools with its games, and by 1999 had a dedicated department to feed designers and producers working on new projects with

57. Why does BMW post a toolkit on its website?

同义转述

A) Because it wants to interest more customers.

B) Because it wants to improve their website.

C) Because it wants their customers to give advices or ideas on their products.

D) Because it wants see if the customers' ideas match their prototype.

58. We may conclude from the text that _____.

细节推断

A) EA is a computer game producer

B) EA is the largest hi-tech company in the world

C) "Red Alert" made its first appearance before 1996

D) Westwood Studios used to be owned by EA for many years

59. Which of the following behavior does not reflect that we are now in a customer-driven market？

A) BMW posts a toolkit to collect customers' ideas.

B) GE brings up 25 luminaries to discuss the evolution of GE's technology.

C) Westwood establishes a department to deal with customers' innovations.

综合推断

D) GE's healthcare division calls some of the well-

customer innovations of existing ones. "The fan community has had a tremendous influence on game design", says Mr Verdu, "and the games are better as a result."

Researchers call such customers "lead users". GE's healthcare division calls them "luminaries". They tend to be well-published doctors and research scientists from leading medical institutions, says GE, which brings up to 25 luminaries together at regular medical advisory board sessions to discuss the evolution of GE's technology. GE then shares some of its advanced technology with a subset of luminaries who are from an "inner sanctum of good friends", says Sholom Ackelsberg of GE Healthcare. GE's products then emerge from collaboration with these groups.

published doctors and research scientists "luminaries".

60. Which of the following can replace the word "customer-driven"? (词义理解)

A) customer-centered

B) customer-satisfied

C) customer-analyzed

D) customer-evaluate

61. Customers invited by BMW didn't want any money, instead, they just want _____.

A) to be invited in a conference

细节推断 B) their suggestions and ideas to be accepted by the company and be of use in the cars' upgrade

C) take a look at BMW's newest models

D) get together and exchange experience on driving the BMWs

【短文大意】德国宝马汽车公司开始于两年前在自己的网站上放置了一个工具包，以便利用它的消费者的创造力。这个工具包让宝马的消费者就公司如何充分利用先进的技术和在线服务表达他们的想法。宝马汽车公司从1000 名登陆该工具包的消费者中选了 15 名，并邀请他们去见慕尼黑的工程师。德国宝马汽车公司认为他们的一些建议（一些现在仍在保密中）已达到样板标准的水平。"他们很高兴被我们邀请，我们的技术专家对他们的想法很感兴趣，"Mr Reimann 说，"他们不想要任何报酬。"

Westwood 工作室，是美国艺电公司拥有的一个游戏开发工作室，在 1996 年推出游戏 "Red Alert" 后，首先关注它的消费者对产品的改革和创新的想法：游戏者对现有的游戏提出了新要求，并把它免费上传到游戏迷网站上。Westwood 做了一个明智的决定，就是要利用游戏者改革和创新的想法。很快，该公司就运用了基础游戏开发工具，到 1999 年又成立了一个专门的部门利用消费者对现有游戏的改革和创新想法来制定计划进行设计和生产。"游戏迷社区对游戏设计产生了很大的影响"，Mr Verdu 说"游戏最终会更完美"。

研究人员将这些消费者称为"领先使用者"。通用电气公司的保健部门叫他们"智囊团"。GE 说，他们将会成为来自主要医疗机构中公认的好医生和研究科学家，这些主要医疗机构培养了 25 名才智出众的人一起参与日常医学顾问董事会议，共同讨论 GE 通用电气公司的技术发展问题。通用电气公司的保健专家 Sholom Ackelsberg 说，通用电气公司之后得与智囊团，也就是来自"内部的好朋友们"共同分享它在技术上取得的进步。通用电气公司的产品是在与智囊团合作之后诞生的。

57.【题目译文】为什么宝马在自己的网站上放置了一个工具包？

【答案及解析】C) 根据第一段所陈述，工具包的作用是让 BMW 的用户在汽车改进等多方面提供意见和建议，因此 C)为正确答案。其余的三项都没有指明工具包的作用，因而也不能说明 BMW 发布工具包的目的。

58.【题目译文】我们可以从这篇文章中推断出_____。

【答案及解析】A) 见文中第二段第一句话 "Westwood Studios, a game developer now owned by EA, …" 可以判断 A) EA 是电脑游戏生产商为正确选项。文中没有提到 EA 是全球最大的游戏生产商，因此排除 B)。C)也不符合文中内容"Red Alert" in 1996。D)陈述 Westwood Studios 属于 EA 已多年，也没有在文章中体现，故排除 D)。

59.【题目译文】下面哪种行为没有反映出我们现在正处于以客户为中心的市场？

【答案及解析】D) 选项 A)、B)、C)分别是文中出现的三家公司用以收集客户意见的方式和渠道，体现了我们正处于一个以客户为中心的市场中，因而排除 A)、B)、C)项，D)为正确选项。

60.【题目译文】下面哪个词可以替换 customer-driven？

【答案及解析】A) A)项是"以客户为中心"的意思；B)项是"让客户满意"的意思；C) 项是"客户分析"的意思；D)项是"客户评估"的意思。本文分别以三家公司用以收集客户意见的方式和渠道来改进和发展自己的产品和技术为例，体现了我们正处于一个以客户为中心的市场中，故 A) "以客户为中心"为正确选项。

61.【题目译文】被宝马邀请的消费者不想要报酬，而他们仅仅想要_____？

【答案及解析】B) 只有 B)项可以体现以客户为中心的市场中企业和客户间的互动过程。其余三项都不符合题目要求。

Passage Two

Where do pesticides fit into the picture of environmental disease? We have seen that they now pollute soil, water, and food, that they have the power to make our streams fishless and our gardens and woodlands silent and birdless. Man, however much he may like to pretend the contrary, is part of nature. Can he escape a pollution that is now so thoroughly distributed throughout our world ?

We know that even single exposures to these chemicals, if the amount is large enough, can cause extremely severe poisoning. But this is not the major problem. The sudden illness or death of farmers, farm workers, and others exposed to sufficient quantities of pesticides are very sad and should not occur. For the population as a whole, we must be more concerned with the delayed effects of absorbing small amounts of the pesticides that invisibly pollute our world.

Responsible public health officials have pointed out that the biological effects of chemicals are cumulative over long periods of time, and that the danger to the individual may

62. Which of the following is closest in meaning to the sentence "Man, however much he may like to pretend the contrary, is part of nature."?

A) Man appears indifferent to what happens in nature.

B) Man can escape his responsibilities for environmental protection.

C) Man acts as if he does not belong to nature.

D) Man can avoid the effects of environmental pollution.

63. What is the author's attitude towards the environmental effects of pesticides?

A) Indifferent   B) Pessimistic

C) Concerned   D) Defensive

64. In the author's view, the sudden death caused by exposure to large amounts of pesticides _____.

A) now occurs most frequently among all accidental deaths

B) is not the worst of the negative consequences resulting from the use of pesticides

C) has sharply increased so as to become the center of public attention

D) is unavoidable because people can't do without pesticides in farming

65. People tend to ignore the delayed effects of exposure to chemicals because _____.

A) limited exposure to them doesn't do much harm to people's health

B) the present is more important for them than the future

69

depend on the sum of the exposures received throughout his lifetime. For these very reasons the danger is easily ignored. It is human nature to shake off what may seem to us a threat of future disaster. "Men are naturally most impressed by diseases which have obvious signs", says a wise physician, Dr. Rene Dubos, "yet some of their worst enemies slowly approach them unnoticed."

细节推断

C) the danger does not become apparent immediately

D) humans are able to withstand small amounts of poisoning.

66. It can be concluded from Dr Dubo's remarks that _____.

A) people find invisible diseases difficult to deal with

B) people tend to overlook hidden dangers caused by pesticides

C) diseases with obvious signs are easy to cure

细节推断

D) attacks by hidden enemies tend to be fatal

【短文大意】在环境已经被破坏后杀虫剂还有适用的地方吗？我们已看到目前杀虫剂污染了土壤、水和食物，我们也清楚杀虫剂有强大的力量使我们的溪水里的鱼儿灭绝，使我们的花园和森林里的鸟死光。人类，尽管常常表现得恰好相反，实际上却是自然界的一部分。人类能逃过今天已经遍布全世界的污染吗？

我们知道甚至仅仅接触化学药品，如果剂量很大，也可以产生极端严重的毒害。但这不是主要的原因。农民突然生病或死亡，农场工人和其他人员接触大量的杀虫剂都是很糟糕的事和不应该发生的事。从总体上说对于人类，我们必须更关心因吸收了少量的不容易被察觉的，污染我们世界的杀虫剂所产生的延迟后果。

负责公众健康的官员已指出化学药品的生物影响是长时间积累的，对一个人的危险可能是由他一生中吸入的化学药品的总量决定的。由于这些原因，危险是很容易被忽视的。人类的本性就是忽视那些看起来可能对我们将来构成威胁的灾难。"人们自然地对那些有明显症状的疾病有印象"，一个有见识的内科医生 Dr Rene Dubos 说，"然而一些最危险的敌人在慢慢地不知不觉地接近他们。"

62.【题目译文】下列哪个句子最接近 "Man, however much he may like to pretend the contrary, is part of nature." 这句话的意思？

【答案及解析】C) 语义理解题。答案关键在于正确理解 "like to pretend the contrary" 几个词的含义，它的字面含义为：喜欢假装相反。通过上下文，这一句的意思是：人类，尽管常常表现得恰好相反，却是自然的一部分。比较四个选项可知 C)项（人类表现得好像他不属于自然）是正确答案。

63.【题目译文】哪个是作者对杀虫剂对环境产生影响的态度？

【答案及解析】C) 本题就作者对杀虫剂会造成的环境后果的态度提问。从第一段最后一句的反问来看，作者忧心忡忡。作者在文中一一描述了杀虫剂明显和潜在的两方面危险，提醒人们重视其潜在的危害，可见作者对这一问题是关注的，故而 C)是正确答案。

64.【题目译文】根据作者的观点，接触大量杀虫剂导致突然的死亡_____。

【答案及解析】B) 本题答案主要见于第二段。可以从 "But this is not the major problem."和"For the population as a whole, we must be more concerned with the delayed effects …" 中推断作者认为由于大剂量吸入杀虫剂而突然致死还不是最糟糕的，我们还应更关注因吸收少量的杀虫剂所产生的延迟后果和危害，因而 B)是正确答案。

65.【题目译文】人们忽视化学药品带来的延迟后果是因为_____。

【答案及解析】C) 见原文最后一段 "Men are naturally most impressed by diseases which have obvious signs,"，人们自然地对那些有明显症状的疾病有印象，言下之意，忽视潜伏危险的原因是这种危险不会马上显现，故而 C)是正确答案。

66.【题目译文】从 Dr Dubo 的话中可以推断出_____。

【答案及解析】D) 从文中最后一句话中的" yet some of their worst enemies slowly approach them unnoticed."（然而一些最糟的敌人慢慢地不知不觉地接近他们。）可以推断出潜伏的敌人才是最危险的，所以 D) 为正确选项。

Part V   Cloze

【短文大意】今天，汽车是美国最流行的一种交通工具。它已经作为一种交通方式完全地取代了马。美国人在处理个人事务上百分之九十使用他们的车。大多数美国人能够买得起车。一台正规生产的车在 1950 年是 50 元，1960 年是 470 元，到了 1975 年则上升到了 750 元。

在这期间，美国汽车生产商开始改进产品，提高效率。因此，从 1950 年至 1975 年，每个家庭的年收入增长速度比汽车价格上涨的速度快。因此在今天，购买一台新车的花费占据一个家庭总收入的份额就相对较小了。在 1951 年，从比例上说，购买一台新车会花费每个家庭 8.1 个月的工资。而在 1975 年，它只需要 4.75 个月的工资。除此之外，1975 年生产的车在技术上要优于以往的车。机动车的影响已经蔓延至整个经济，因为车是如此的重要，美国人在保管车上要比在其他项目上的花费多得多。

67.【答案】C 本句义为：汽车取代了马，成为日常交通工具。replace "取代"；deny "否定、否认"；reproduce "复制"；ridicule "嘲笑"。故选 replace。

68.【答案】A 句义见上题。means 指 "工具"，复数形式作单数理解，其他几项不符合句义。

69.【答案】B nearly "接近"；hardly "几乎不"；certainly "肯定"；somehow "设法"。接近 90% 符合句义。

70.【答案】A personal "个人的"；personnel "人事的"；manual "手工的"；artificial "人造的"。本句义为：美国人在处理个人事情上 90% 会用到车。

71.【答案】A 根据上下文，本句的意思应该是 "大部分的美国人能够买车"。

72.【答案】B 根据上下文，这里应该是指 "正规" 的生产。

73.【答案】B 指的是在 1975 年，与前面的 in 1950 对应。

74.【答案】D 本句的意思是 "汽车生产商开始改进产品，提高效率"，只有 improve "改进" 符合句义。raise 表示 "提高"，make 表示 "制造"，reduce 表示 "减少"。

75.【答案】C 根据上下文，只能选 average "平均"。unusual 表示 "不同寻常的"。

76.【答案】C 本句的意思是 "家庭收入的增长速度比汽车价格上涨的速度快"。

77.【答案】D 本句缺主语，只有 A 和 D 可作主语，但 A 项的意思不符。purchasing "购买"，符合句义。

78.【答案】A 本句指的是家庭收入较小的一部分，不是指具体的数目。

79.【答案】B 此处需要用副词，先排除 percentage 这个名词，其他选项中只有 proportionally "相应、成适当比例的" 符合句义。本句句义是 "在不同年代车价与家庭收入按比例增加"，其他两项意思不符。

80.【答案】A 本句指的是家庭收入，只有 income 符合句义。

81.【答案】C 在表示某物花费某人多少钱的意思时，若主语为物，动词要用 cost。

82.【答案】A 本句指的是 4.75 个月的收入。

83.【答案】B 本句义为在 1975 年出产的汽车在工艺上比前些年出产的各种型号的汽车优越。be superior to "比……优越"；be better 后面要接 than；famous "著名的"；fastest "最快的"。

84.【答案】C 本句意为汽车的影响涉及整个经济。influence "影响"；affect 是动词，不符合语法；running 和

notice 不符合句义。

85.【答案】B as 在这里引导原因状语从句。

86.【答案】B spend (in) doing sth.是固定搭配。

Part VI   Translation

87. The rent is reasonable and ,_____（此外）, the location is perfect.

   【题目译文】这个租金很合理，此外，地段很好。

   【答案及解析】moreover 注意副词 moreover 表示补充的含义。

88. Driving again after his accident must_____(需要很大勇气).

   【题目译文】事故后再次驾车需要很大勇气。

   【答案及解析】have taken a lot of courage 注意 "courage" 为不可数名词。

89. Miss Smith was _____(处于最佳状态) when she played the piano.

   【题目译文】当史密斯女士演奏钢琴时，她处于最佳状态。

   【答案及解析】at her best   at one's best: 处于最佳状态。

90. They can not choose but_____(承认他们错了).

   【题目译文】他们只好承认他们错了。

   【答案及解析】admit they are wrong can not…, but do sth.：只得做什么事。

91. He is an artist _____(而不是一名教师).

   【题目译文】他是一位艺术家，而不是一名老师。

   【答案及解析】rather than a teacher   短语 rather than 而不

大学英语4级考试新题型模拟试卷五答案及详解

参考答案

Part I   Writing

Why is Imported Fast-Food So Popular?

In recent years, leading fast-food restaurant chains from abroad have been doing booming business in China. In the commercial centers of all the major cities，you cannot help noticing their attractive neon signs and advertisements for delicious snacks. McDonald's, Kentucky Fried Chicken and other fast-food outlets have captured the imagination especially of young people in China, and they have become some of the most fashionable places to eat nowadays.

It can be said that there are two keys to their success: a "hard" one and a "soft" one . The former is their attention to good sanitation and quality control. The latter is their effort to appeal to children and young people by arranging parties with free toys and gifts.

Some people criticize these foreign restaurants, saying that Chinese food is healthier than the hamburgers, etc. This may be true, but we must ask ourselves why Chinese fast-food restaurants have failed to take off. Perhaps if we learn both the "soft" and the "hard" secrets of success from the newcomers，we can give a boost to the domestic catering industry.

Part II   Reading Comprehension (Skimming and Scanning)

1. Y   2. N   3. NG   4. N   5. Y   6. Y   7. N

8. abstract thinking   9. the United States   10. Chandigarh

Part III   Listening Comprehension

Section A

11. A   12. D   13. A   14. D   15. B   16. A   17. A   18. B   19. A   20. B
21. D   22. D   23. D   24. C   25. C

Section B

26. C   27. B   28. D   29. A   30. B   31. B   32. C   33. B   34. D   35. D

Section C

36. means   37. services   38. aims   39. arouse

40. persuade   41. media   42. press   43. commercial

44. that the growth in advertising is one of the most striking features of the western world

45. whether the cost of advertising is paid for by the manufacturer or by the customer

46. It is clear that it is the customer who pays for advertising

Part IV   Reading Comprehension (Reading in Depth)

Section A

47. G: purposes   48. C: include   49. A: each   50. E: preparations   51. O: size
52. J: available   53. H: obtain   54. K: especially   55. N: deadly   56. B: clean

Section B

57. D   58. B   59. C   60. C   61. A   62. A   63. B   64. D   65. C   66. A

Part V   Cloze

67. B   68. D   69. C   70. A   71. C   72. B   73. D   74. C   75. C   76. B

77. A   78. D   79. B   80. C   81. D   82. B   83. C   84. B   85. D   86. B

Part VI   Translation

87. along with alcohol / in addition to alcohol   88. what's more he ate a lot

89. attributes his success to   90. First of all   91. So far as I'm aware/ As far as I know

答案解析及录音原文

Part I   Writing

Why is Imported Fast-Food So Popular?

In recent years, leading fast-food restaurant chains from abroad have been doing booming business in China. In the commercial centers of all the major cities you cannot help noticing their attractive neon signs and advertisements for delicious snacks. McDonald's, Kentucky Fried Chicken and other fast-food outlets have captured the imagination especially of young people in China, and they have become some of the most fashionable places to eat nowadays.

It can be said that there are two keys to their success: a "hard" one and a "soft" one . The former is their attention to good sanitation and quality control. The latter is their effort to appeal to children and young people by arranging parties with free toys and gifts.

Some people criticize these foreign restaurants, saying that Chinese food is healthier than the hamburgers, etc. This may be true, but we must ask ourselves why Chinese fast-food restaurants have failed to take off. Perhaps if we learn both the "soft" and the "hard" secrets of success from the newcomers we can give a boost to the domestic catering industry.

【评论】本文是一篇议论论文，文章开头提纲挈领，阐明了要议论的话题。中间一段对出现的现象做出了明确的解释。结尾处表达出一些人对此的看法。这篇作文层次分明，逻辑性强。

本文的一个闪光之处是恰当地使用了一些常用句型：

It can be said that…,   The former is…,   the latter is…,   This may be true,   but we must do sth.,   Perhaps if we do sth…

Part II   Reading Comprehension(Skimming and Scanning)

【文章及答案解析】本文讲述的是不同的文化对空间概念有不同的理解，文章在首段提到了西方文化的空间概念，从而引出其他的文化特点，并进行了对比。

The Cultural Patterning of Space   【文章大意】   空间的文化模式

Like time, space is perceived differently in different cultures. Spatial consciousness in many Western cultures is based on a perception of objects in space, rather than of space itself. Westerners perceive shapes and dimensions, in which space is a realm of light, color, sight, and touch. Benjamin L. Whorf, in his classic work Language, Thought and Reality, offers the following

主旨大意   1. The passage is about cross-cultural spatial perceptions. (Y)

细节推理   2. European cultures generally value inner personal experience more than non-European cultures do. (N)

explanation as one reason why westerners perceive space in this manner. Western thought and language mainly developed from the Roman, Latin-speaking, culture, which was a practical, experience-based system. Western culture has generally followed Roman thought patterns in viewing objective "reality" as the foundation for subjective or "inner" experience. It was only when the intellectually crude Roman culture became influenced by the abstract thinking of the Greek culture that the Latin language developed a significant vocabulary of abstract, nonspatial terms. But the early Roman-Latin element of spatial consciousness, of concreteness, has been maintained in Western thought and language patterns, even though the Greek capacity for abstract thinking and expression was also inherited.

However, some cultural-linguistic systems developed in the opposite direction, that is, from an abstract and subjective vocabulary to a more concrete one. For example, Whorf tells us that in the Hopi language the word heart, a concrete term, can be shown to be a late formation from the abstract terms think or remember. Similarly, although it seems to Westerners, and especially to Americans, that objective, tangible "reality" must precede any subjective or inner experience; in fact, many Asian and other non-European cultures view inner experience as the basis for one's perceptions of physical reality. Thus although Americans are taught to perceive and react to the arrangement of objects in space and to think of space as being "wasted" unless it is filled with objects, the Japanese are trained to give meaning to space itself and to value "empty" space.

It is not only the East and the West that are different in their patterning of space. We can also see cross-cultural varieties in spatial perception when we look at arrangements of urban space in different Western cultures. For instance, in the United States, cities are usually laid out along a grid, with the axes generally north/south and east/west. Streets and buildings are numbered sequentially. This arrangement, of course,

像时间一样,不同的文化对空间的感知是不同的。西方许多国家对空间的意识基于对物体在空间中的感觉,而不是对空间本身的感觉。西方人对形状和维度的感觉,就是空间是一种光线、颜色、视觉和触觉的领域。本杰明·沃尔夫在他的经典著作《语言、思维和现实》中,就用下面的解释来说明西方人为什么会以这种方式来感觉空间。西方的思维和语言主要源于以拉丁语为母语的古罗马文化,而这是一种基于实际

细节确认

8. Ancient Greek culture emphasized _____.
(**abstract thinking**)

和经验的体系。一般来说,西方文化已经采用古罗马人的思维模式,把客观"实体"视为主观或者内在经验的基础。一直到这种在知性上不成熟的古罗马文化受到希腊文化的抽象思维的影响时,拉丁语言才发展出一套意义重大的词汇——抽象的非空间术语。但是空间意识和具体化的古罗马—拉丁成分已经在西方思维和语言模式中保存了下来,尽管也继承了希腊人的抽象思维和表达能力。

然而,还有些文化语言系统朝着相反的方向发展,就是从一套抽象、主观的词汇发展到一套更为具体的词汇。例如,沃尔夫告诉我们,在霍皮语中,"心"这个字是一个具体的术语,可它是在先有了"思维"和"记忆"这种抽象术语之后才形成的。同样地,尽管在西方人,特别是美国人看来,客观的、有形的"实体"一定要先于主观的或者内在经验,但实际上,许多亚洲和非欧洲文化把内在经验看成是对有形实体的感觉的基础。因此,尽管美国人被教导在空间中感知物体的排列和做出反应,会认为除非空间中充满物体,否则就是"被浪费了",而日本人却被教育对空间本身赋予意义,对"空旷"的空间赋予价值。

不仅东西方在空间模式上存有差异,而且当我们用不同的西方文化来观察城市的布局时,还可以凭对空间的感觉体会交叉文化的不同。例如,在美国,城市的布局通常是沿着一个网格展开,轴心一般是南北向和东西向。街道和建筑物按顺序编号。当然这种安

9. Streets and buildings are numbered sequentially in _____. (**the United States**)

75

makes perfect sense to Americans. When Americans walk in a city like Paris, which is laid out with the main streets radiating from centers, they often get lost. Furthermore, streets in Paris are named, not numbered, and the names often change after a few blocks. It is amazing to Americans how anyone gets around, yet Parisians seem to do well. Edward Hall, in The Silent Language, suggests that the layout of space characteristic of French cities is only one aspect of the theme of centralization that characterizesFrench culture. Thus Paris is the center of France, French government and educational systems are highly centralized, and in French offices the most important person has his or her desk in the middle of the office.

Another aspect of the cultural patterning of space concerns the functions of spaces. In middleclass America, specific spaces are designated for specific activities. Any intrusion of one activity into a space that it was not designed for is immediately felt as inappropriate. In contrast, in Japan, this case is not true: Walls are movable, and rooms are used for one purpose during the day and another purpose in the evening and at night. In India there is yet another culturally patterned use of space. The function of space in India, both in public and in private places, is connected with concepts of superiority and inferiority. In Indian cities, villages, and even within the home, certain spaces are designated as polluted, or inferior, because of the activities that take place there and the kinds of people who use such spaces. Spaces in India are segregated so that high caste and lowcaste, males and females, secular and sacred activities are kept apart. This pattern has been used for thousands of years, as demonstrated by the archaeological evidence uncovered in ancient Indian cities. It is a remarkably persistent pattern, even in modern India, where public transportation reserves a separate space for women. For example, Chandigarh is a modern Indian city designed by a French architect. The apartments were built according to European concepts, but the Indians living

排对美国人来说是完美的。当美国人在像巴黎这样的城市漫步时，他们往往会迷路。因为巴黎的街道是从中央发散开来的。此外，巴黎的街道是命名的而不是按序编号的，而且常常不用经过几个街区，街名就变换了。美国人对当地人如何到处行走大为疑惑，而巴黎人却显得行动自如。霍尔在《无声的语言》一书中指出：法国城市空间布局的特点仅仅是反映法国文化特征中中央集权的一个方面。因此巴黎是法国的中心，法国政府和教育系统高度集中。在法国人的办公室里，最重要的人物的办公桌位于办公室的中央。

细节确认

3. China is an example of a highly centralized society. (NG)

空间文化模式的另一个方面涉及到空间的各种功能。在美国的中产阶层，特定的空间是为特定的活动而设计的。任何活动，一旦跨越其特定空间，人们立刻就会觉得不合事宜。比较之下，在日本就不是那样。墙壁可以移动，房间的使用目的在白天和晚上是不一样的。在印度，又是另一种空间使用的文化模式。印度的公共和私人场所在功能上均有优劣的概念。在印度的城市、乡村甚至是家庭里，某些场所因为所从

细节确认

4. Japan and the United States are similar in that both cultures use the same space for a variety of different purposes. (N)

事的活动和使用这些场所的人的缘故而被认定是肮脏或者卑劣的。印度的空间是隔离开来的，以便社会

同义转述

5. In India, public and private space is separated for males and females. (Y)

等级高的和等级低的、男的和女的、世俗的和神圣的活动都分开进行。这种模式沿用了几千年，在印度古城挖掘出来的考古证据就说明了这一点。即便在现代的印度，这种空间模式仍旧相当地清晰和顽固，哪怕是在公共交通车辆上也要把妇女使用的空间隔离开来。例如，昌迪加尔是印度的一座由法国建筑师设计的现代化城市。公寓大楼都是按欧洲理念建造的，但

there found certain aspects inconsistent with their previous use of living space. Ruth Freed, an anthropologist who worked in India, found that Indian families living in Chandigarh modified their apartments by using curtains to separate the men's and women's spaces. The families also continued to eat in the kitchen, a traditional pattern, and the living room-dining room was only used when Western guests were present. Traditional Indian village living takes place in an area surrounded by a wall. The courtyard gives privacy to each residence group. Chandigarh apartments, however, were built with large windows, reflecting the European value of light and sun, so many Chandigarh families pasted paper over the windows to recreate the privacy of the traditional courtyard. Freed suggests that these traditional Indian patterns may represent an adaptation to a densely populated environment.

Anthropologists studying various cultures as a whole have seen a connection in the way they view both time and space. For example, as we have seen, Americans look on time without activity as "wasted" and space without objects as "wasted". Once again, the Hopi present an interesting contrast. In the English language, any noun for a location or a space may be used on its own and given its own characteristics without any reference being made to another location or space. For example, we can say in English: "The room is big" or "The north of the United States has cold winters." We do not need to indicate that "room" or "north" has a relationship to any other word of space or location. But in Hopi, locations or regions of space can not function by themselves in a sentence. The Hopi can not say "north" by itself; they must say "in the north", "from the north", or in some other way use a directional suffix with the word north. In the same way, the Hopi language does not have a single word that can be translated as room. The Hopi word for room is a stem, a portion of a word, that means "house", room" or "enclosed chamber", but the stem can not be used alone. It must be joined to a suffix

是住在那里的印度人却发现某些方面与他们以前居住的空间模式不一致。在印度工作的人类学家鲁思·弗里德发现，居住在昌迪加尔城的许多印度家庭都改造了他们的公寓，用窗帘把男人和女人的空间隔离开来。

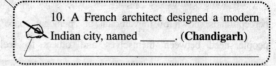

10. A French architect designed a modern Indian city, named _____. (Chandigarh)

只有自家人时，他们就仍然依照传统模式在厨房里吃饭，而有西方客人光临时他们才启用客厅或是饭厅。传统的乡村都生活在由墙围绕的区域内，院子给每户人家提供了隐私的空间。然而，昌迪加尔城的公寓大楼，建有很多敞的窗户，从而折射出欧洲人对光线与阳光的重视。而许多昌迪加尔城的家庭却把窗户玻璃上糊满了纸张以便重建传统的院子里的隐私空间。弗里德认为这些传统的印度模式也许反映出人们对人口密集型环境的一种适应。

从整体上研究不同文化的人类学家已经察看到了时间观与空间观之间的联系。例如，正如我们所察看到的，美国人把没有活动的时间看作是"被浪费了的时间"，把没有物体的空间看作是"被浪费了的空间"。霍皮人再一次提供了有趣的对比。在英语中，任何表示地点或者空间的词都可以单独使用，能呈现出各自的特征而无需任何参照。例如，在英语里，可以说："这房间很大"或者"美国北方的冬天很寒冷"。我们无需表明"房间"或者"北方"与任何其他表空间或地点的词语有联系。但在霍皮语里，地点或者空间地域的词语本身不能在句子里单独使用。霍皮人不能单独地使用"北方"这个词，他们得说"在北方"、"从北方"或者用另一种方式给"北方"这个词加上一个方向性的后缀。同样地，霍皮语没有一个单词能够被翻译成"房间"。霍皮语中的"房间"是词干，是意思为"房屋"、"房间"或"居室"词的一部分，但是不能单独使用，必须加上后缀才使这个词表示"在房子里"或"从居室"。霍皮语中像"房间"、"居室"或"大厅"这些表示空洞空间的概念只有跟其他空间关联时才具有意义。

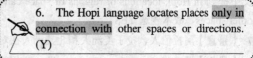

6. The Hopi language locates places only in connection with other spaces or directions. (Y)

that will make the word mean "in a house" or "from a chamber". Hollow spaces like room, chamber, or hall in Hopi are concepts that are meaningful only in relation to other spaces.

In some cultures a significant aspect of spatial perception is shown by the amount of "personal space" people need between themselves and others to feel comfortable and not crowded. North Americans, for instance, seem to require about four feet of space between themselves and people near them to feel comfortable. On the other hand, people from Arab countries and Latin America feel comfortable when they are close to each other. People from different cultures, therefore, may unconsciously infringe on each other's sense of space. Thus just as different perceptions of time may create cultural conflicts, so too may different perceptions of space.

细节
推理

7. Arab, Latin American, and North American cultures all have similar perceptions of personal space. (**N**)

在某些文化中空间感觉的一个重要方面就是通过人们所需的彼此感觉舒适却又不觉拥挤的"私人空间"体现出来的。例如，北美人彼此感觉舒适所需的空间距离大约是 4 英尺。而阿拉伯人和拉美人反而是彼此靠近才会感觉舒服。因此，不同文化的人可能会无意间侵犯别人的空间感。正如不同的时间观可能会造成文化上的冲突，不同的空间观也会引发同样的问题。

【答案解析】

1.【题意】本文讲述的是不同文化对空间的感知。

【解析】本题考查的是文章的主旨，常用的句式是 The passage is (mainly) about....。文章的主题在首句得到了说明：Like time, space is perceived differently in different cultures. 不同的文化对空间的概念不同。接下来文章对其进行了进一步的说明。different cultures 与 cross-cultural 同义，故本题选 YES。

2.【题意】欧洲文化一般对个人内在价值的评价多于非欧洲文化。

【解析】文章首段提到了西方文化的空间概念以 a perception of objects in space 为基础，具有 practical 和 experience-based 的特点，将 objective reality 作为 inner experience 的基础。而第二段以 however 开始，引出其他文化的特点。本段提到 many Asian and other non-Europe cultures 将 inner experience 作为基础。题干与原文相悖，故此题选 NO。

3.【题意】中国是高度集中的国家代表。

【解析】文章第三段谈到空间布局时指出法国是一个高度集中的国家，但没有提到中国，故此题选 NG。

4.【题意】日本和美国的文化相似点都是为了各种各样不同的目的而使用相同的空间。

【解析】根据题干中谈的 space 的 purpose 可以将查找的内容锁定在涉及 function of spaces 的第四段，purpose 与 function 同义。该段指出 In contrast, in Japan, this case is not true，这说明美国与日本截然不同，故此题选 N。

5.【题意】在印度，公共和私人空间是男性和女性分用的。

【解析】根据题干快速寻找关于 India 的部分。第四段提到了 function of space in India，接着说明 Spaces in India are segregated... males and females... kept apart，说明空间男女分用。故本题选 YES。

6.【题意】霍皮语指定的地点只与其他的空间或方向有关。

【解析】题干中提到 Hopi 以及 connection with other spaces 等。第五段首句提到 connection，接着查找 Hopi，文中提到 In Hopi, locations or regions of space cannot function by themselves in a sentence，而是必须 in relation to other spaces，故本题选 YES。

7.【题意】阿拉伯人、拉丁美洲人和北美洲人在对个人空间的感知上都有相似之处。

【解析】题干与 perception of personal space 相关，对应着最后一段首句 spatial perception... personal space. 根据

on the other hand 可以判断 North Americans 与 people countries and Latin America 在此问题上不同，故本题选 NO。

8.【题意】古希腊文化强调抽象思维。

【解析】在文章中，只有首段提到了 Greek culture，具体的短语有 the abstract thinking of the Greek culture 以及 the Greek capacity for abstract thinking，因此该题的答案是 abstract thinking。

9.【题意】在美国街道和建筑物都按顺序标号。

【解析】在第三段提到了 For instance, in the United States… Streets and building are numbered sequentially，因此本题的答案是 the United States。

10.【题意】由法国建筑师设计的现代印度城市取名为昌迪加尔(Chandigarh)。

【解析】本题和上一个题一样，就是在举例处出题。第四段中间 For example, Chandigarh is… designed by a French architect，题干将被动句换成了主动句，故本题的答案是 Chandigarh。

Part Ⅲ   Listening Comprehension
Section A

11. M: Are you going to replace the light switch by yourself?

    W: Why should I call an electrician?

    Q: What does the woman imply?

【答案及解析】判断推理题。男士问女士是否自己换电灯开关，女士回答说，何必请电工。此处关键是理解 why 的用法，它既可表示建议，也可表示置疑，并非总是表示询问。

12. W: The map shows that this street goes downtown.

    M: Yes, but what we want to know is how to get to the park.

    Q: What does the man mean?

【答案及解析】判断推理题。女士说此地图表明这条街道通往市中心，但男士说此时他们想知道的是怎样去公园。

13. W: Listen, the heating is broken in my room. When are you going to come and fix it?

    M: I'm sorry to hear that, madam. I'll get someone right away.

    Q: What is the probable relationship between the man and the woman?

【答案及解析】判断推理题。女士房间的暖气坏了，想知道什么时候才会有人来修，男士说马上派人来。此处关键是 get someone right away，因此此人代表公司。

14. M: Well, I thought you were working for a large business machine company.

    W: I did for a few months, but I decided that accounting wasn't what I liked best.

    Q: Why did she want to change her job?

【答案及解析】因果判断题。男士以为女士在一家大型商业机构里工作，女士回答确实在那儿工作过几个月，但发现做会计不是她最喜欢的工作。

15. M: I wish to buy a new car, but I spent too much on my house last year.

    W: New cars are expensive. You can't borrow so much money.

    Q: What happens to the man?

【答案及解析】判断推理题。男士想一辆新车，但去年买房花钱太多，女士回答说新车很贵，他不可能借那么多钱。此处关键是 you can't borrow so much money，省略了 from me。

16. M: Well, if I had begun studying English earlier, I wouldn't be having so much trouble with my pronunciation.

79

W: Oh, I can understand why you feel that way.

Q: What problem does the man have?

【答案及解析】虚拟语气。男士说如果早一点学英语,现在就不会有发音问题。女士说能理解他的感受。

17. M: Have you filled out your tax forms yet?

W: Don't remind me of them! They're so confusing that I'm discouraged before I start!

Q: What emotion is the woman feeling?

【答案及解析】语气判断题。男士问女士是否填了税收表,女士说,别提了,太复杂了,还没填就感到泄气。关键词是 discouraged,意为沮丧,泄气。

18. W: Will you come to my novel reading next week?

M: I'll be out of town then.

Q: What does the man mean?

【答案及解析】判断推理题。女士问男士是否会参加她下星期的小说朗读,男士说下星期要出城。男士的回答即为一种间接否定。

Now you will hear two long conversations.

Conversation One

M: That was my mother on the phone. She and my father will be here Friday.

W: Yeah...

M: What's the matter? Don't you like them?

W: Sure I do! It's their smoking I don't like. I really don't want them to smoke in the house.

M: They're just here for the weekend. It's not a big problem, is it?

W: Second-hand smoke is a big problem around here. It's dangerous—especially for the baby. Besides, everything stinks for a week!

M: I know what you mean. But what can we do.

W: We can ask them to smoke outside.

M: I guess you're right. OK. You can tell them when they get here.

W: Me? Are you kidding? They're your parents! You tell them.

Questions 19 to 21 are based on the conversations you have just heard.

19. What was the possible relationship between the speakers?

【解析】推理判断题。通篇文章二人语气非常亲密,并且涉及家庭内部矛盾问题,因此答案是 A)。

20. What happened to the speakers?

【解析】事实细节。首先确定是男士的父母要来,由此排除 A) 和 D),而后又提到男士的父母吸烟使女士为难,故可以排除 C)。

21. What could we infer from the conversations?

【解析】推理判断题。对话中女士一直抱怨男士的父母抽烟所带来的麻烦,可见她讨厌男士的父母抽烟。

Conversation Two

M: Hello, is that Jane?

W: Speaking.

M: Hi, Jane, this is Chris here. Listen, I'm in real trouble. I'm in the middle of an assignment and my computer's crashed!

W. Oh, no! Bad luck.

M: Yeah, I can't believe it! What do you think I should do? I'm desperate!

W: I think I can help you, Chris, so calm down! It happened to me last year, and I solved the problem. So don't panic! What happened exactly?

M: I started to boot up and suddenly the screen went blank. I couldn't do anything! All my data's gone! I can't retrieve it! I've lost it all!

W: Listen, that happened to me, and I took it round to a small company I know and they fixed it in an hour! And they retrieved all my data, too.

M. Thank god! Can you give me the phone number?

W: Sure, they're very helpful. Speak to Kit, Kit Marlow. You can mention my name too ! That might help. Do you have a pen handy? Call them at 0208 346 789. Oh, just one more thing. Before you do that, switch it off, and try again. You never know it might correct itself.

M. Okay! I'll do that right now. And thanks a million!

W: My pleasure. Good luck.

Questions 22 to 25 are based on the conversations you have just heard

22． Where might the conversation take place?

【解析】推理判断题。对话一开始男士问：Hello, is that Jane?女士回答：Speaking.(是我)。这是电话用语，显然两人在打电话。

23． What is the man's trouble?

【解析】事实细节题。男士说遇到大麻烦了，他正在工作时，电脑突然崩溃了。(I'm in the middle of an assignment and my computer's crashed!）

24． Why does the woman think she can help?

【解析】事实细节题。女士说可以帮助男士，因为她去年遇到过这个问题，并成功地解决了。(It happened to me last year, and I solved the problem.)

25． How can the man retrieve his data?

【解析】推理判断题。男士说计算机里所有的数据都丢失了，再也无法恢复了，女士安慰说她的电脑曾发生过同样的问题，她拿到一个小公司去修理，一个小时就修好了，丢失的数据也找回来了。女士的话隐含的意思是，男士也可以让那个小公司帮助恢复数据。

Section B

Passage One

The dog has often been an unselfish friend to man. It is always grateful to its master. It helps man in many ways. Certain breeds of dogs are used in criminal investigations. They are trained to sniff out drugs and bombs. They help police to catch criminals. Some dogs are trained to lead blind people.

The dogs that help in criminal investigations are trained at a school called the Military Dog Studies branch of the US Air Force in Lackland, Texas. The dogs to be trained are selected by an air force team. This team visits large cities across the country to buy the dogs. They may buy dogs from private citizens for up to $750 each. Some citizens freely give their dogs. The dogs selected must be healthy, brave and aggressive. They must be able to fight back if they are attacked. The dogs chosen are between the ages of one and three. They are given a medical examination when they arrive at the school. Their physical examination includes X-rays and heart tests. The trainee dogs undergo the first stage of training when they arrive in Lackland. This is an 11-week course for patrol duty. After this course, the best dogs are selected to go on another 9-week course. They learn drug-sniffing or bomb-sniffing. After this course, the dogs are ready for their jobs in the cities or on air force bases.

The training given to a drug-sniffing dog is different from that given to a bomb-sniffing dog. A drug-sniffing dog is trained to scratch and dig for the drugs when he sniffs them. A bomb-sniffing dog sits down when he finds a bomb. That is the alert for hidden explosives.

Questions 26 to 28 are based on the passage you have just heard.

【内容概要】狗是人类无私的朋友，它们以各种方式帮助人类. 有些狗被用于犯罪调查。一个空军小分队专门负责买狗，这些狗要接受 11 个星期的巡逻训练，表现好的狗将再接受 9 个星期的训练，此间它们要学习嗅毒品或炸弹。

26．How are the dogs obtained for training?

【解析】细节考查题。一个空军小分队专门负责挑选狗，他们到全国各大城市买狗。他们以每只高达 750 美元的价格从私人手里买狗，也有些人将狗白送给他们。

27．What is the first stage of training for the dogs?

【解析】细节考查题。狗要接受两个阶段的训练，第一个阶段为期 11 周，第二阶段为期 9 周。

28．What do dogs learn during the 9-week training?

【解析】细节考查题。在为期 9 周的训练中，它们要学习嗅毒品或炸弹。

Passage Two

From this lookout we enjoy one of the most spectacular views of San Francisco. As you can see, the city rests on a series of hills varying in altitude from sea level to nine hundred and thirty-eight feet. The first permanent settlement was made at this site in 1776. For thirteen years the village had fewer than one hundred inhabitants. But in 1848, with the discovery of gold, the population grew to ten thousand. The same year the name was changed from Yerba Buea to San Francisco. By 1862 telegraph communications linked San Francisco with eastern cities, and by 1869, the first transcontinental railroad connected the Pacific coast with the Atlantic seaboard. Today San Francisco has a population of almost three million. It is the financial center of the west, and serves as the terminus for trans-Pacific steamship lines and air traffic. The port of San Francisco which is almost eighteen miles long with forty-two piers, handles between five and six million tons of cargo annually.

And now, if you look to your right, you should just be able to see the east section of the Golden Gate Bridge. The bridge, which is more than one mile long, spans the harbor from San Francisco to Marin County and the Red Wood Highway. It was completed in 1937 at a cost of thirty-two million dollars and is still one of the largest suspension bridges in the world.

Questions 29 to 31 are based on the passage you have just heard.

【内容概要】本文讲了旧金山的历史。它的人口从最初的不到 100，发展到一万，又到今天的三百万。1862 年电报把旧金山与东部城市联系起来，1869 年第一条洲际铁路把太平洋海岸和大西洋海岸连接起来，目前它是西部的金融中心。旧金山的金门桥把该港口延伸到马林县及红林高速公路。该桥建于 1937 年，耗资三千二百万，仍为世界上最大的吊桥之一。

29．According to the tour guide, what happened in 1848?

【解析】细节考查题。1848 年随着金子的发现，人口增长到一万。

30．What is the population of San Francisco today?

【解析】细节考查题。今天旧金山的人口约为三百万。

31．How much did it cost to complete the construction of the Golden Gate Bridge?

【解析】细节考查题。该桥建于 1937 年，耗资三千二百万，仍为世界上最大的吊桥之一。

Passage Three

For good or bad, computers are now part of our daily lives. With the price of a small home computer now being lower, experts predict that before long all schools and businesses and most families in the rich parts of the world will own a computer of some kind. Among the general public, computers arouse strong feelings——people either love them or hate them. The computer lovers talk about how useful computers can be in business, in education and in the home——apart from all the games, you can do your accounts on them, use them to control your central heating, and in some places even do your shopping with them. Computers, they say, will also bring some leisure, as more and more unpleasant jobs are taken over by computerized robots.

The haters, on the other hand, argue that computers bring not leisure but unemployment. They worry, too, that people who spend all their time talking to computers will forget how to talk to each other. And anyway, they ask, what's wrong with going shopping and learning languages in classroom with real teachers? But their biggest fear is that computers may eventually take over from human beings altogether.

Questions 32 to 35 are based on the passage you have just heard.

【内容概要】电脑已成为我们生活的一部分。然而，电脑在普通大众中引起强烈的反应——人们要么爱要么恨。电脑爱好者会提到电脑的诸多好处，恨者则有诸多担心。

32．What does this passage mainly talk about?

【解析】推理判断题。电脑在普通大众中引起强烈的反应——人们要么爱，要么恨。

33．According to the passage, what is not mentioned about computers?

【解析】文中提到电脑对做生意、对教育及家庭非常有用，甚至可以代替人类做一些对人不利的工作，但它可能导致失业。

34．What is the biggest fear of the computer haters?

【解析】细节考查题。电脑憎恨者的最大担心是电脑可能最终完全代替人类。

35．What's the speaker's attitude to computers?

【解析】推理判断题。文章既提到电脑爱好者的态度，又提到电脑憎恨者的态度，用词客观，不带任何个人感情。

Section C

Advertising can be thought of as the (36) means of making something known in order to buy or sell goods or (37) services. Advertising (38) aims to increase people's awareness and (39) arouse interest. It tries to inform and to (40) persuade. The (41) media are all used to spread the message. The (42) press offers a fairly cheap method. The cinema and (43) commercial radio are useful for local markets. Television, although more expensive, can be very effective. There can be no doubt (44) that the growth in advertising is one of the most striking features of the western world. We might ask (45) whether the cost of advertising is paid for by the manufacturer or by the customer, since advertising forms part of the cost of production, which has to be covered by the selling price. (46) It is clear that it is the customer who pays for advertising.

Part IV   Reading Comprehension (Reading in Depth)

Section A

【短文大意】我们从哪儿、怎样获得安全的饮用水呢？通常情况下你需要一加仑的水，这仅是为了满足每人每天饮用的需求。这仅仅是用于饮用，不包括做饭用水、洗澡、浪费的水或是用在宠物身上的水。一条狗，你每天得另外需要一加仑的水；一只猫，一天（你需要）大约是一品脱的水。

对于那些在灾难来临时毫无准备的人来说，还可以从几处获得水。如果你能够及时想到的话，每个卫生间上层水槽装有几加仑的水，可以抽出来饮用。你的热水器，依据大小的不同，还会有30到50加仑的水，你可以从底部的水龙头把水放出来。冰箱里的冰块也是可用水。你可能还可以从家中的所有管道中抽出更多加仑的水，以供饮用，这取决于房子的情况。

如果你已经用尽了从可靠来源获得的水，那该怎么办？特别是在灾难时刻，假设没有储备水或购买的水均被污染了，那么即使一股清澈的水流都关乎生死。

如果你找到的水是乌黑、浑浊的，首先，用数层纸巾、干净的布片或咖啡过滤器过滤出污垢，然后用一种方法或别的方法净化它。

47.【答案及解析】G) 所填词与 drinking 一起作 for 的介词宾语，for 表示目的，与此意义一致的名词为 purposes。

48.【答案及解析】C) 所填词在 does not 之后，为及物动词，接宾语 water，表示 this 与 water for cooking…的关系。根据前面的 for drinking only 可知，这里所说的水量只表示饮用水，不包括做饭等用水。因此选择 include 与 not 连用，将后面的内容排除。

49.【答案及解析】A) 所填词修饰 dog，结合后面的 and for a cat，选用 each，构成并列关系，并与上句的 per person 对应。

50.【答案及解析】E) 所填词是 made 的宾语，为名词。由下文的种种方法可见，这里谈的是未做准备（preparations）时的情况。而 efforts 虽然也能和 made 搭配，但没有概括出本段的意思。

51.【答案及解析】O) 空格处需要填入名词，而 depending on 说明所填词与句子其他部分（某个变量）有很大的联系。根据 thirty to fifty gallons 可以判断，水的多少与容量大小相关，选择 size。

52.【答案及解析】J) 第二段主要谈了水源的获取，所填词修饰 water，说明冰箱中的冰块也是可用（available）的水。这里一直在谈水源的获取，而非洁净与否，排除 clean，注意从篇章的角度进行选择。

53.【答案及解析】H) 介词 to 后缺一个动词，动词宾语为 more…water，有关水的获取选用 obtain。

54.【答案及解析】K) 根据给出的选项，只有副词符合句法的要求。选项中 especially 强调了灾难发生时的情形，而 physically 与 a time of disaster 无法连贯。

55.【答案及解析】N) 系动词 be 后缺少形容词。本句中 even 表明句意是对上一句的递进，故选择意义上比具有消极含义的 polluted 更进一步的 deadly。

56.【答案及解析】B) 最后两句话谈到了水的净化（purify）问题，选择形容词 clean 修饰 cloth。

Section B

Passage one

I'm usually fairly skeptical about any research that concludes that people are either happier or unhappier or more or less certain of themselves than they were 50 years ago. While any of these statements might be true, they are practically impossible to prove scientifically. Still, I was struck by a report which concluded that

语义理解

57. The author thinks that the conclusions of any research about people's state of mind are_____.

A) reasonable

B) confusing

C) illogical

D) questionable

today's children are significantly more anxious than children in the 1950s. In fact, the analysis showed, normal children aged 9 to 17 exhibited a higher level of anxiety today than children who were treated for mental illness 50 years ago.

Why are America's kids so stressed? The report cites two main causes: increasing physical isolation—brought on by high divorce rates and less involvement in community, among other things—and a growing perception that the world is a more dangerous place.

Given that we can't turn the clock back, we can still do plenty to help the next generation cope.

At the top of the list is *nurturing*(培育)a better appreciation of the limits of individualism. No child is an island. Strengthening social ties helps build communities and protects individuals against stress.

To help kids build stronger connections with others, you can pull the plug on TVs and computers. Your family will thank you later. They will have more time for face-to-face relationships, and they will get more sleep.

Limit the amount of *virtual* (虚拟的) violence your children are exposed to. It's not just video games and movies; children can see a lot of murder and crime on the local news.

Keep your expectations for your children reasonable. Many highly successful people never attended Harvard or Yale.

Make exercise part of your daily routine. It will help you cope with your own anxieties and provide a good model for your kids. Sometimes anxiety is unavoidable. But it doesn't have to ruin your life.

58. What does the author mean when he says, we can't turn the clock back?

细节推断

A) It's impossible to slow down the pace of change.

B) The social reality children are facing can not be changed.

C) Lessons learned from the past should not be forgotten.

D) It's impossible to forget the past.

59. According to an analysis, compared with normal children today, children treated as mentally ill 50 years ago_____.

A) were less isolated physically

B) were probably less self-centered

同义转述 C) probably suffered less from anxiety

D) were considered less individualistic

60. The first and most important thing parents should do to help their children is_____.

A) to provide them with a safer environment

B) to lower their expectations for them.

同义转述 C) to get them more involved socially.

D) to set a good model for them to follow.

61. What conclusion can be drawn from the passage?

同义转述 A) Anxiety, though unavoidable, can be coped with.

B) Children's anxiety has been enormously exaggerated.

C) Children's anxiety can be eliminated with more parental care.

D) Anxiety, if properly controlled, may help children become mature.

【短文大意】现在的人与50年前的人相比，是更快乐还是更不快乐，是更自信还是更不自信，我对此的研究成果表示怀疑。然而，这些陈述可能都是对的，只是我们无法科学地加以证明。但我始终对一项报告记忆犹新：现在的孩子明显比20世纪50年代的孩子更焦虑。实际上，分析表明，现在正常的9~17岁的孩子比50年前被视为患有精神疾病的孩子要焦虑得多。

为什么美国的孩子现在如此焦虑呢？报告中列举了两个主要原因：一是由于离婚率上升和接触社会的机会减少而导致孩子的孤独；二是他们觉得这个世界越来越没有安全感。

即使我们不能让时光倒流，我们还是能做很多事情来帮助下一代。

首要的是让他们认识到个人力量的局限性。没有哪一个孩子是孤岛。加强他们与社会的联系有助于建立集体并能使他们尽量远离焦虑。

你可以通过拔掉电视和电脑的电源插座来加强孩子与他人的联系，全家人都会因此而感激你。这样孩子们之间可以有更多面对面相互交流的机会，他们也可以得到更多的睡眠。

减少孩子面对虚拟暴力的机会。虚拟暴力不仅指录像、游戏和电影，而且还指孩子看到的当地新闻中许多有关谋杀及犯罪的报道。

对孩子的期望要理智。许多非常成功的人士都没有进过哈佛大学和耶鲁大学。

让运动成为你每天必做的事情。它会有助于减少你自己的焦虑，同时你可以给孩子树立良好的榜样。有时，焦虑是不可避免的，但不会毁了你的生活。

57.【题目译文】作者认为关于人精神状况的任何研究成果是_____。

【答案及解析】D) 根据文中第一句和第二句可知：作者认为现在关于人们精神状况的一些研究是值得怀疑的，也就是说作者认为是有问题的，因此 D)为正确答案。

58.【题目译文】当作者说"我们不能让时间倒流"时，作者的意思是_____。

【答案及解析】B) 50 年以来，美国的儿童面对越来越多的压力，也就是发生了儿童所面对的种种家庭和社会的问题。他说时间无法倒流，真正的意图就是我们无法改变儿童所面对的社会现实，因此 B)为正确答案。

59.【题目译文】根据分析，与现在的正常孩子相比，50 年前被视为患有精神疾病的孩子_____。

【答案及解析】C) 可以从第一段最后一句 "In fact, the analysis showed, normal children . . . than children who were treated for mental illness 50 years ago." 找到答案，即分析表明，现在正常的 9~17 岁的孩子比 50 年前被视为患有精神疾病的孩子要焦虑得多。故而可以推断 C)为答案。

60.【题目译文】家长应该帮助孩子做的最主要也是最重要的事情是_____。

【答案及解析】C) 从第四段中 "At the top of the list is nurturing a better appreciation of the limits of individualism." 和 "Strengthening social ties helps build communities and protects individuals against stress." 可以确定 C)为答案。即：首要的是让他们认识到个人力量的局限性。同时加强他们与社会的联系有助于建立集体并使他们尽量远离焦虑。因此父母可以帮助孩子加强与社会的更多联系。

61.【题目译文】根据文章能够得出什么推断？

【答案及解析】A) 较难取舍的是 A）和 D）。D）项说焦虑如果能正确控制的话，可以帮助孩子们变得成熟。我们从整篇文章中都找不到能让孩子们变得成熟的词句来，故应选 A）。尽管焦虑是不可避免的，但不会毁了一个人的生活，文章建议可以通过很多方法和途径来减少焦虑。这道题有一定的难度。

Passage Two

Whether the eyes are "the windows of the soul" is debatable; that they are intensely important in interpersonal communication is a fact. During the first two months of a baby's life, the stimulus that produces a smile is a pair of eyes. The eyes need not be real: a mask with two dots will produce a smile. Significantly, a real human face with eyes covered will not motivate a smile, nor will the sight of only one eye when the face is presented in profile. This attraction to eyes as opposed to the nose or month continues as the baby matures. In one

同义转述

接63B

62. The author is convinced that the eyes are_____.

A) of extreme importance in expressing feelings and exchanging ideas

B) something through which one can see a person's inner world

C) of considerable significance in making conversations interesting

D) something the value of which is largely a matter of long debate

study, when American four-year-olds were asked to draw people, 75 percent of them drew people with mouths, but 99 percent of them drew people with eyes. In Japan, however, where babies are carried on their mother's back, infants do not acquire as much attachment to eyes as they do in other countries. As a result, Japanese adults make little use of the face either to encode (把……编码) or decode (解码) meaning. In fact, Argyle reveals that the "proper place to focus one's gaze during a conversation in Japan is on the neck of one's conversation partner".

The role of eye contact in a conversational exchange between two Americans is well defined: speakers make contact with the eyes of their listener for about one second, then glance away as they talk, in a few moments they reestablish eye contact with the listener to reassure themselves that their audience is still attentive, then shift their gaze away once more. Listener, meanwhile, keep their eyes on the face of the speaker, allowing themselves to glance away only briefly. It is important they be looking at the speaker at the precise moment when the speaker reestablishes eye contact: if they are not looking, the speaker assumes that they are disinterested and either will pause until eye contact is resumed or will terminate the conversation. Just how critical this eye maneuvering is to the maintenance of conversational flow becomes evident when two speakers are wearing dark glasses: there may be a sort of traffic jam of words caused by interruption, false starts, and unpredictable pauses.

63. Babies will not be stimulated to smile by a person_____.
A) whose front view is fully perceived
同义转述 B) whose face is seen from the side
C) whose face is covered with a mask
D) whose face is free of any covering

64. According to the passage, the Japanese fix their gaze on their conversation partner's neck because_____.
A) they don't like to keep their eyes on the face of the speaker
B) they need not communicate through eye contact
C) they don't think it polite to have eye contact
同义转述 D) they didn't have much opportunity to communicate through eye contact in babyhood

65. According to the passage, a conversation between two Americans may break down due to_____.
A) one temporarily glancing away from the other
同义转述 B) eye contact of more than one second
C) improperly-timed ceasing of eye contact
D) constant adjustment of eye contact

66. To keep a conversation flowing smoothly, it is better for the participants_____.
细节推理 A) not to wear dark spectacles
B) not to make any interruptions
C) not to glance away from each other
D) not to make unpredictable pauses

【短文大意】眼睛是否是"心灵的窗口"的说法是有争议性的；眼睛在人与人的交往沟通中是极为重要的，这是事实。在婴儿最初的两个月期间，使它们微笑的刺激物正是一双眼睛。不必是一双真眼睛：有两个小圆点的面具便能使之微笑。有意思的是，盖住眼睛的真人脸却不能让它微笑，脸侧对着孩子的人也不会引起婴儿发笑。眼睛的吸引力而非鼻子或嘴的吸引力一直持续到孩子长大。在一次研究中，当让4岁的美国儿童画人时，他们中75%画的是带嘴巴的人，而99%的孩子画的是带眼睛的人。然而在日本，婴儿是被背在妈妈后背上的，他们很少像其他国家婴儿那样对眼睛有太多的感情。结果是，日本的成人很少用脸去表达含义。事实上，Argyle透露说，在日本，谈话期间目光放在对方的脖子上才是合适的位置。

眼睛交流的作用在两个美国人交谈中被很好地定义了：说话者与听者的眼睛接触大约一秒钟，之后边交谈目光边转移到别的地方，一会儿再次与听者目光接触以确定对方在听自己讲话，之后再一次转移目光。同时，听者的目光也注视着说话者的脸，他们的目光只允许短时间地转移。当说话者再次与听者目光接触时听者也同样在注视着说话者，这是非常重要的：假如这时候听者没有注视着说话者，说话者会猜想他们对自己的讲话不感兴趣，他们要么停下来直到听者的目光再一次转移到自己身上，要么中断谈话。这种眼睛的巧妙移动可以保证说话畅通，其重要性在两个人都带着墨镜谈话时变得很明显：谈话不顺畅可能是因中断谈话、不正确的开场白，以及不可预料的停顿而造成的。

62.【题目译文】作者深信眼睛是_____。

【答案及解析】A) 第一段第一句话 "Whether the eyes are the 'windows of the soul' is debatable; that they are intensely important in interpersonal communication is a fact." 即"眼睛是否是'心灵的窗口'的说法是有争议性的；眼睛在人与人的交往沟通中是极为重要的，这是事实。" A) 眼睛在表达情感和交流思想方面非常重要与第一句话内容相吻合，因为情感表达和思想交流是人际交往中非常重要的一部分。

63.【题目译文】婴儿不会被_____的人刺激而微笑。

【答案及解析】B) 根据文章第一段中的第二、第三和第四句话，A、C、D 中所提到的几种人都可能逗婴儿笑。B 项内容与第四句后半句"脸侧对着孩子的人不会引婴儿发笑"相符，故正确。

64.【题目译文】根据文章，日本人把他们的目光集中在其谈话对象的脖子上是因为_____。

【答案及解析】D) 见第一段第七句 "In Japan, however, where babies are carried on their mother's back, infants do not acquire as much attachment to eyes as they do in other countries." 然而在日本，婴儿是被背在妈妈后背上的，他们很少像其他国家婴儿那样对眼睛有太多的感情。结果是，日本的成人很少用脸去表达含义。日本人把他们的目光集中在其谈话对象的脖子上，故 D)正确。

65.【题目译文】根据文章，两个美国人之间的谈话可能终止是由于_____。

【答案及解析】C) 见第二段第二、三句。可以确定 C)为答案。因为说话者发现听者没有注视他，说话者会猜想他对自己的讲话没兴趣，他们要么停下来直到听者的目光再一次转移到自己身上，要么中断谈话。因此两个美国人之间的谈话可能终止是由于不恰当地终止眼睛交流造成的。

66.【题目译文】为保证谈话顺利进行，谈话参与者最好要_____。

【答案及解析】A) 见文章最后一句，"Just how critical this eye maneuvering is to the maintenance of conversational flow becomes evident when two speakers are wearing dark glasses: there may. . . . " 因为谈话者都带着墨镜，这阻碍眼睛交流的顺利进行，故而谈话就不会顺畅， 因此 A)为正确答案。

Part V　　Cloze

【翻译】在各种建立起来的语言里，有两类词构成了整个词汇表。第一种单词是我们在日常对话中所熟悉的词汇，也就是说，这类词我们是从我们自己的家庭成员中或者从我们熟识的人中所学会的。就算我们不会读也不会写，我们也他们能够了解并使用。

这些词涉及到生活中的普通物品，是所有使用它的人相互交流的词汇。这类词可以称之为大众词汇。因为他们是大多数人所使用的并且不是某一个特定阶层所独享的。从另一方面讲，我们的语言还包括另一种由许多相对而言在日常交流当中不常用的词汇。每一位受过教育的人都知道他们的意思，但是在家中或者在市场却没有必要用到它们。这类词汇，我们最初既不是从母亲的口中也不是从同学那里了解的，而是从我们所读过的书中、从我们参加的讲座中，或者是从受过高等教育者的对话中所学习到的。这些受到过高等教育的人讨论问题的方式超越了日常生活的范围。这些单词可以称之为"学术性词语"，这两种词汇的差别以及"大众性词语"是我们理解语言的一个重要的过程。

67.【答案】B 本句中由 with which 引导的定语从句，修饰先行词 those words。短语 become acquainted with sb./sth.意为"认识某人，了解某事。"

68.【答案】 D imitate 意为"模仿", stimulate 意为"刺激，激发"。study 和 learn 都有"学习"的意思，study

着重研究，而 learn 指一般性的学习，故选 D。

69.【答案】C mate 意为"伙伴，同事"，可组成复合名词，如：classmate 同学，roommate 同房间的人。relative 意为"亲戚"，member 意为"成员"，family member 意为"家庭成员"，fellow 意为"伙伴，家伙"。

70.【答案】A which 引导非限制性定语从句，和前面的定语从句并列，修饰先行词 those words，关系代词 that 只能在限制性定语从句中代替 which，故选 A。

71.【答案】C even if 在这里引导条件状语从句。even 是副词，不能引导状语从句。in spite of 和 despite 表示"尽管"，为介词词组或介词，也不能引导状语从句。

72.【答案】B 本句的意思是"它们涉及生活中的一般性事情，是所有使用这种语言的人惯用的语言材料。concern 意为"涉及"，mind 和 care 表示"介意，计较"，relate 表示"讲述、叙述"。

73.【答案】D use 意为"使用"，apply 意为"运用"，hire 意为"雇用"，adopt 意为"采纳"。本文要表达的意思是"使用"，所以 D 正确。

74.【答案】C at large 意为"普遍的、一般的"，in public 意为"公开地、当众"，at most 意为"至多、不超过"，at best 意为"充其量、至多"。

75.【答案】C share 意为"份额、共享"。right 和 privilege 意为"权利、特权"，放在本句中不符合题意。possession 意为"拥有、占有"，通常指拥有财物。

76.【答案】B comprise "包含、包括、由……组成"。compose 常用于被动结构 be composed of 表示"由……组成"。consist 是不及物动词，必须和 of 组成短语动词表示"由……组成"，constitute 意为"构成"。

77.【答案】A seldom 意为"不经常、很少"。

78.【答案】D prospect 意为"前景"；way "方式"；reason "理由"；necessity "必要性"。本句只有 necessity 符合句意。

79.【答案】B 本句意为"我们最初既不是从母亲嘴里，也不是从同学那里了解这些单词的……" first "第一、首先"；primary "基本的、原始的"；prior "优先的、在先的"；principal "主要的、首要的"。

80.【答案】C learn sth. from one's lips 是固定搭配，表示"从某人嘴里得知"。

81.【答案】D but 在这里表示转折的含义。

82.【答案】B attend a lecture "参加一个讲座"。

83.【答案】C formal "正式的"；former "以前的"；formula "公式、方程"；formative "形成的"。

84.【答案】B topic "话题"；theme "主题"；point "要点"。本句指讨论的话题，故选 topic。

85.【答案】D degree 和 extent 均可表示程度，但 extent 还可表示"范围"。本句意为：……讨论问题的方式超越了日常生活的范围，所以选 extent。border "边界"，link "连接"。

86.【答案】B diversion "转移、转向"；distinction "差别"；diversity "多样性、变化"；similary "相似之处"。本句意为：学术性词语和大众化词语之间的差别，故选 distinction。

Part VI Translation

87. Tobacco is taxed in most countries, _____(除了酒之外).

【题目译文】除了酒之外，香烟在大多数国家都收税。

【答案及解析】along with alcohol / in addition to alcohol 注意介词短语"along with"和"in addition to"的用法。

88. He enjoyed the meal and _____(而且把它全吃光了).

【题目译文】他喜欢吃肉，而且把它全吃光了。

【答案及解析】what's more he ate a lot "what's more"：而且。

89. He _____(把成功归因于) hard work and a bit of luck.

【题目译文】他把成功归因于努力工作和运气。

【答案及解析】attributes his success to "attribute...to": 把...归因于。

90. _____ (首先), I'd like to welcome you to the meeting.

【题目译文】首先，欢迎你能来参加会议。

【答案及解析】First of all. 注意 "First of all" 的用法。

91. _____(就我所知), this is the first time a British rider has won the competition.

【题目译文】就我所知，他是第一位获胜的英国骑士。

【答案及解析】So far as I'm aware/ As far as I know 注意 "so far as" 和 "as far as"：就……的程度，到……程度

大学英语4级考试新题型模拟试卷六答案及详解

参考答案

Part I   Writing

Internet in China

With the arrival of the knowledge economy age, internet, as a new media, has come into our life. Nowadays, getting on the line and using internet has become popular in China, especially in some big cities. Why is internet getting so popular in such a short time?

The popularity of internet results from its great utility. With internet, it is convenient and fast for people to get a variety of information and what they need is just to press the key buttons. E-mail is another popular use. It can transmit your letter quickly, safely and accurately. Besides, Internet can have some other uses. Doctors may use it to diagnose and treat their patients by discussing or exchanging experience with doctors in other parts of the world. Students may obtain new knowledge from a national-long–distance-educational-system via Internet. Finally, businessmen can conduct E-commerce or E-business on the Net.

With so many users and advantages, there is no doubt more and more people will be wired to internet—the information expressway. In short, with its high speed and efficiency, Internet will be more popular in China in the near future.

Part II   Reading Comprehension (Skimming and Scanning)

1. N   2. Y   3. Y   4. Y   5. N   6. Y   7. NG

8. their potential for harm   9. monitored at every dental visit

10. ice packs and non-prescription pain medication

Part III   Listening Comprehension

Section A

11. C	12. D	13. D	14. D	15. C	16. C	17. A	18.A
19. B	20.C	21. D	22. B	23. D	24. C	25. C	

Section B

26. D	27. C	28.A	29. B	30.C	31. C	32.B	33.A	34.B	35.C

Section C

36. surrounded   37. Scottish   38. Although   39. thought

40. visitors   41. printed   42. creature   43. considered

44. Accounts of the Loch Ness monster also sounded like jokes

45. The body was thick in the middle but thin at the tail

46. The monster seemed shy

Part IV   Reading Comprehension (Reading in Depth)

Section A

47. N: closest   48. G: weather   49. L: survive   50. A: height   51. E: extinct

52. K: different   53. B: breathe   54. O: active   55. D: directly   56. M: eat

Section B

57. D	58. A	59. D	60. B	61. B	62. C	63. C	64. B	65. B	66. B

91

Part V   Error Correction

67. ease→easy   68. began→begin   69. later→late

70. consist ^ anything→ of   71. nice→nicely   72. are→is

73. present→presents   74. present→presented   75. and→or   76. how→what

Part VI   Translation

77. On average   78. compared with mine / in comparison with mine

79. did he charge me too much / did he overcharge me

80. in all walks of life   81. dropped out of

答案解析及录音原文

Part I   Writing

Internet in China

With the arrival of the knowledge economy age, internet, as a new media, has come into our life. Nowadays, getting on the line and using internet has become popular in China, especially in some big cities. Why is internet getting so popular in such a short time?

The popularity of internet results from its great utility. With internet, it is convenient and fast for people to get a variety of information and what they need is just to press the key buttons. E-mail is another popular use. It can transmit your letter quickly, safely and accurately. Besides, Internet can have some other uses. Doctors may use it to diagnose and treat their patients by discussing or exchanging experience with doctors in other parts of the world. Students may obtain new knowledge from a national—long distance educational system via Internet. Finally, businessmen can conduct E-commerce or E-business on the Net.

With so many users and advantages, there is no doubt more and more people will be wired to internet—the information expressway. In short, with its high speed and efficiency, Internet will be more popular in China in the near future.

【评论】本文第一段引出文章的话题：因特网的发展。然后在第二段对其发展的原因和用途做了具体的说明。最后一段对因特网在中国的发展寄予了很高的期望。本文层次清晰，语言生动，非常有说服力。

【常用句型】

第一段：With the arrival of, getting on the line and using Internet has become popular in China

第二段：It is convenient and fast for people to get a variety of information

第三段：With so many users and advantages, there is no doubt more and more people will be wired to Internet

Part II   Reading comprehension (Skimming and Scanning)

【文章及答案解析】本文介绍了一些与智齿相关的医学知识。由于在智齿是否一定要拔除这一问题上还存在着很多争议，因此文章没有一概而论，而是就各种具体情况进行了分析。主要从四个层次来进行说明。一是可以依据哪些症状来判断应当拔除智齿；二是拔除智齿的最佳时间；三是涉及拔除的一些有利与不利的方面；最后是拔除智齿后的护理注意事项。最后作者指出，无论拔除与否，都应当在与牙医认真讨论之后做出决定。即使决定不拔除，也应在每次问诊时进行检查。

Wisdom Teeth

Wisdom teeth are the last teeth to erupt. This occurs usually between the ages of 17 and 25. There remains a great deal of controversy regarding whether or not these teeth need to be removed. It is generally suggested that teeth that remain completely buried or un-erupted in a normal position are unlikely to cause harm. However, a tooth may become impacted due to lack of space and its eruption is therefore prevented by gum(齿龈), bone, another tooth or all three. If these impacted teeth are in an abnormal position, their potential for harm should be assessed.

What are the Indications for Removing Wisdom Teeth?

Wisdom teeth generally cause problems when they erupt partially through the gum. The most common reasons for removing them are:

Decay

Saliva, bacteria and food particles can collect around and impacted wisdom tooth, causing it, or the next tooth to decay. It is very difficult to remove such decay. Pain and infection will usually follow.

Gum Infection

When a wisdom tooth is partially erupted, food and bacteria collect under the gum causing a local infection. This may result in bad breath, pain, and swelling. The infection can spread to involve the cheek and neck. Once the initial episode occurs, each subsequent attack becomes more frequent and more severe.

Pressure Pain

Pain may also come from the pressure of the erupting wisdom tooth against other teeth. In some cases this pressure may cause the erosion of these teeth.

智齿是萌出最晚的牙齿，通常在 17～25 岁时长出。关于智齿是否需要拔除还存在很多的争论。医生通常提示：保留完全埋藏在或未萌发在正常位置的牙齿是没有损害的，然而牙齿也可能由于空间不足而阻生，最终在齿龈、骨骼及其他牙齿或全部因素的阻碍下萌发。如果这些阻生的牙齿错位萌出，那么就应该考虑它们的潜在危害。

细节确认

1. Dentists have reached an agreement that wisdom teeth should be removed in case it leads to other problems.（**N**）

8. If the position of the impacted teeth is abnormal, it is advisable to weigh____. (**their potential for harm**)

什么情况下该拔出智齿?

当智齿部分地从牙龈中萌出时，通常会引发很多问题。拔出它们的最常见理由是：

腐蚀

唾液、细菌及食物的残渣会堆积在阻生智齿的周围，引起该牙齿或临近的牙齿腐蚀。这样的腐蚀牙齿是很难拔出的，常伴有疼痛和感染。

牙龈感染

智齿部分萌出时，滞留在牙龈下的食物和细菌会引发局部感染，导致口臭、疼痛和肿胀。炎症能够扩散到包括面部和颈部的地方。一旦开始发作，每次后果都会变得更加频繁和严重。

压痛

智齿萌发时对其他牙齿的挤压也会引起疼痛，有时可能会造成牙齿的腐蚀。

When is the Best Time to Have Wisdom Teeth Removed?

It is now recommended by specialists that impacted wisdom teeth be removed between the ages of 14 and 22 whether they are causing problems or not. Surgery is technically easier and patients recover much more quickly when they are younger. What is a relatively minor operation at 20 can become quite difficult in patients over 40. Also the risk of complications increases with age and the healing process is slower.

Should a Wisdom Tooth be Removed When an Acute Infection is Present?

Generally, no. Surgery in the presence of infection can cause infection to spread and become more serious. Firstly, the infection must be controlled by local oral *hygiene*(卫生) and *antibiotics* (抗生素).

The Pro's and Con's of Wisdom Tooth Removal Some PRO'S Removing a Wisdom Tooth

Wisdom teeth may be hard to access with your toothbrush. Over time, the accumulation of bacteria, sugars and acids may cause a cavity to form in the tooth. If it is restored with a filling, the *cavity*(洞) may spread and destroy more tooth structure and cause severe consequences to the tooth and surrounding supportive structure.

Due to the difficulty of keeping these teeth clean with your daily home care, bacteria and food debris remaining on the wisdom teeth may present a foul smell-causing bad breath.

A wisdom tooth that is still under the gums in a horizontal position (rather than a vertical position) may exert pressure to the surrounding teeth, causing crowding and crooked teeth. This also may occur if there is not enough space in the mouth for the wisdom tooth. This may warrant braces to repair the damage.

拔出智齿的最佳时期

专家们建议无论智齿是否引发问题，人们都应该在14～22岁时将阻生的智齿拔除。手术从技术上来看比较简单，病人越年轻恢复得越快。20岁时所做的小手术相对于40多岁的人来说则变得相当困难，产生并发症的危险也会随着年龄的增长而增加，痊愈的过程也更慢。

> 同义转述
>
> 2. The best time to remove the impacted wisdom teeth is between the ages of 14 and 22 because surgery is technically easier and patients recover much more quickly. (**Y**)

急性感染时应拔除智齿吗？

一般来说，不可以。感染时做手术会造成感染扩散，使病情更加严重。首先，应通过保持局部口腔卫生和使用抗生素来抑制感染。

> 细节确认
>
> 3. If there is an infection, the first thing to do is to stop it before get the wisdom tooth removed. (**Y**)

拔除智齿的利与弊　拔除智齿的理由

智齿很难用牙刷清理干净。时间一长，累积的细菌、糖分及酸性物质就会腐蚀牙齿从而形成牙洞。如果不用填充物填补，牙洞会扩大而且会损坏更多的牙齿，给牙齿及其周围的支撑结构带来严重的后果。

由于日常家庭护理很难保持牙齿的清洁，残留在智齿上的细菌和食物就会产生恶臭带来不良的口气。

埋于牙龈里的水平阻生（不是垂直阻生）智齿会使周围的牙齿受到压力，造成牙齿的拥挤和扭曲，也可能会使智齿在口腔中没有足够的空间生长。这就需要用畸牙矫正架修复损伤。

> 同义转述
>
> 4. Wisdom tooth may become the cause of bad breath. （**Y**）

Some Con's of Removing the Wisdom Teeth:

Depending on the size shape and position of the tooth, removal can vary from a simple extraction to a more complex extraction. With a simple extraction, there is usually little swelling and/or bleeding. More complex extraction will require special treatment which may result in more bruising, swelling and bleeding. However, your dental professional will provide you with post treatment instructions to minimize these side effects.

Following an extraction, a condition called "dry socket" may occur. If the blood *clot*(凝结)that formed in the extraction area becomes removed, it exposes the underlying bone. This condition is very painful, but resolves after a few days. It is preventable by following the post treatment instructions provided by your dental professional.

The longer you wait and the older you get, there is the potential for more problems to occur. This is because as you get older, the bone surrounding the tooth becomes denser, making the tooth more difficult to remove. The healing process may also be slower.

Post Operative Care

Do Not Disturb the Wound

In doing so you may invite irritation, infection and/or bleeding. Chew on the opposite side for the first 24 hours.

Do Not Smoke for 12 Hours

Smoking will promote bleeding and interfere with healing.

Do Not Spit or Suck Through a Straw.

This will promote bleeding and may remove the blood clot, which could result in a dry socket.

Control of Bleeding

If the area is not closed with stitches, a pressure pack made of folded sterile (消毒的) gauze(纱布) pads will be placed over the socket. It is important that this pack stay in place to control bleeding and to encourage

不应该拔除智齿的理由

根据牙齿的体积形状和生长位置的不同，智齿的拔除可以由简单到复杂。简单的摘除，常伴有轻微的肿胀和出血。略微复杂的摘除，则需要可能会导致青肿和出血的特殊处理。但牙医会为你进行后期的治疗指导，将其负面影响减至最小。

细节确认

5. Removing the wisdom teeth involve very complex extraction. （**N**）

拔除后的智齿，可能会引发"干槽症"（牙槽窝骨髓炎）。牙槽内形成的血块开始脱落，骨壁暴露，碰触时十分疼痛，但几天后就可痊愈。专业牙医在进行后期治疗指导时，可以防止此情况的发生。

拖延拔除的时间越大，年龄越大，由此带来的潜在问题就越多。这是由于随着年龄的增大，牙齿周围的骨质密度越来越高，智齿更加巩固而不容易拔除，愈合的过程也就越慢。

术后护理

不要碰触伤口

在护理的过程中伤口可能会受到刺激、感染或出血。24 小时内应用另一侧牙齿咀嚼。

12 小时内不允许吸烟

吸烟将引发伤口出血以及影响愈合。

不要吐唾液或通过吸管吸食

这些都将导致由于出血和血块脱落而引发的干槽症。

同义转述

6. After your wisdom tooth is removed, you'd better not spit or suck through a straw. (**Y**)

止血

如果拔出的部位不需要用线缝合，应在牙槽上放置一块无菌纱布卷。该纱布卷对止血和促进血块凝结

clot formation. The gauze is usually kept in place for 30 minutes. If the bleeding has not stopped once the original pack is removed, place a new gauze pad over the extraction site.

Control of Swelling

After surgery, some swelling is to be expected. This can be controlled by using the cold packs, which slow the circulation. A cold pack is usually placed at the site of swelling during the first 24 hours in a cycle of 20 minutes on and 20 minutes off. After the first 24 hours, it is advisable to rinse(漱口) with warm saltwater every two hours to promote healing. (one teaspoon of salt to eight ounces of warm water).

Medication for Pain Control

Pills such as Aspirin can be used to control minor discomfort following oral surgery. Stronger medicines may be prescribed by the dentist if the patient is in extreme discomfort.

Diet and Nutrition

A soft diet may be prescribed for the patient for a few days following surgery.

Following the removal of your wisdom teeth it is important that you call your dentist if any unusual bleeding, swelling or pain occurs. The first 6～8 hoursafter the extraction are typically the worst, but are manageable with ice packs and non-prescription pain medication. You should also plan to see your dentist approximately one week later to ensure everything is healing well.

It is very important to talk to your dentist about extraction procedure, risks, possible complications and outcomes of the removal of these teeth. The actual extraction may be done by a dentist or it may be referred to an oral surgeon, who is a specialist. This decision is based on the dentist's preference and the unique features of each individual case.

是很重要的。纱布卷通常置于伤口处约 30 分钟，一旦拿掉纱布后仍有出血现象，可另置一块于拔除处。

细节确认

7. After your wisdom tooth is removed, you should rinse with warm water every morning for two weeks. （NG）

止胀

手术后，拔除的部位会有一些肿胀。可以用冰袋冷敷，以减缓血液循环的速度。通常是在术后 24 小时内每隔 20 分钟轮流将其放在肿胀的部位。24 小时后，每隔 2 小时用温盐水漱口以促进伤口的愈合。（一茶匙的盐对八盎司的温水）

止痛的药物

术后可使用如阿司匹林之类的药物减缓轻微的不适。如果病情严重，可用医生开的有较强功效的药物进行治疗。

饮食和营养

术后的几天应吃柔软的食物。

拔除智齿后最重要的是一旦发现有异常的出血、肿胀或疼痛，应和医生联系。术后最初的 6～8 小时内会很严重，可以用冰敷或服用非处方镇痛剂来缓解。约一周后与医生联系以确保一切愈合得很好。

与医生谈论有关手术的过程、具备的风险、可能产生的并发症以及术后的结果是非常重要的。实际上，手术可能由一名牙医完成或者托付给作为专家的口腔外科医生。这一决定取决于牙医的选择和每个病例的独特性。

10. The painful time during the first 6～8 hours after the removal of the tooth can be controlled with_____.(ice packs and non-prescription pain medication)

如果你不确定是否继续进行牙医给的专业性治疗建议，最好的办法是再次咨询。即使咨询后你决定不拔除任何牙齿，你也应该在每次问诊时进行检查。

If you are unsure about whether or not to proceed with the treatment suggested by your dental professional, it is a good idea to get a second opinion. If you decide after consulting with a dentist not to have any teeth extracted, they should be monitored at every dental visit

9. Even if you have decided not to get your wisdom teeth removed, you still have to have them_____. (**monitored at every dental visit**)

【答案解析】

1.**【题意】**智齿都应当拔除，以免引起其他问题，牙医们在这一点上已达成共识。

【解析】本句的关键词是 reached an agreement, wisdom teeth should be removed，由此定位到第一段，"There remains a great deal of controversy regarding whether or not these teeth need to be removed." 就智齿是否拔除的问题仍有大量的争议，而没有达成共识。可见本题应选 NO。

2.**【题意】**拔除智齿的最佳时间是 14～22 岁之间，因为从技术上讲要容易一些，病人康复得也会更快。

【解析】本题的关键是 The best time to remove, between the ages of 14 and 22. 文章小标题 "最佳拔除时间" 部分提到以上的原因。并且在 "术后护理" 前面的一段也提到，如果拖延拔除时间，随着年龄的增长，牙齿周围的骨骼结构会更紧密，拔除也就更困难了，所以应选 YES。

3.**【题意】**如果口腔有感染，拔除智齿首要是消除感染。

【解析】本题的关键词是 infection，原文中提到，如口腔有感染还拔除智齿只会导致感染范围更大、更严重，所以应先使用局部口腔清洁和抗生素解决感染的问题，故应选 YES。

4.**【题意】**智齿会导致口臭。

【解析】本题的关键词是 bad breath，由此定位到 "拔除智齿的理由" 部分，"Due to the difficulty of keeping these teeth clean with your daily home care, bacteria and food debris remaining on the wisdom teeth may present a foul smell-causing bad breath." 因为智齿可能生长在牙刷很难触到的地方，时间久了会堆积细菌和食物残留，所以产生不良口气，故选 YES。

5.**【题意】**拔除智齿要涉及很复杂的过程。

【解析】本题的关键词是 Removing, complex extraction，由此寻读到 "不应该拔除智齿的理由" 部分，原文说 "With a simple extraction, there is usually little swelling and/or bleeding. More complex extraction will require special treatment." 可见题干信息与原文有出入，本题的答案为 NO。

6.**【题意】**智齿拔除后最好不要吐唾液，也不要用吸管喝饮料。

【解析】本题的关键词是 not spit or suck through a straw，可定位到 "术后护理" 部分，原文提到 "Do Not Spit or Suck Through a Straw"，智齿拔除后如果吐唾液或用吸管喝饮料，并会冲走血凝块，导致口腔干燥，所以应选 YES。

7.**【题意】**智齿拔除后应每天早晨用温盐水漱口。

【解析】本题的关键词是 rinse with warm water，答案在 "术后护理" 部分："After the first 24 hours, it is advisable to rinse with warm saltwater every two hours to promote healing (one teaspoon of salt to eight ounces of warm water)." 拔除智齿 24 小时后才可以每隔两小时用温盐水漱口，并且要注意盐的用量，但是没有提及要持续两周。题干信息原文没有指明，故选 NG。

8.**【题意】**如果受阻牙齿的位置很不正常，最好是要权衡一下它潜在危害。

【解析】本题的关键词是 "impacted teeth、abnormal、advisable，" 解在本文的第一段，本段就很清楚地指出："However, a tooth may become impacted due to lack of space and its eruption…If these impacted teeth are in an abnormal position, their potential for harm should be assessed." 可见这里要填的是 their potential for harm.

9. 【题意】即使你决定不要拔除智齿，但在每次看牙医时还是要跟踪观察。

【解析】本题的关键词是"decided not、removed，"解在最后一段，"If you decide after consulting with a dentist not to have any teeth extracted, they should be monitored at every dental visit." 与医生协商决定不拔除后，仍要保持密切的观察。

10. 【题意】拔除智齿后的6～8小时可以通过冰袋和非处方止痛药来度过这段痛苦时间。

【解析】本题的关键词是 first 6～8 hours. 故本题定位到倒数第三段中"The first 6～8 hours after the extraction are typically the worst, but are manageable with ice packs and non-prescription pain medication." 一般拔除智齿后的6～8小时是最难过的，但可以用冰袋和非处方止痛药来控制。大约一星期后可以再去看牙医确保恢复正常。

Part Ⅲ   Listening   Comprehension

Section A

11. M: How many students passed the College English Test last term?

W: Well, let me see, 1016 students took the exam, but half of them failed.

Q: How many students did the woman believe had passed the exam?

【答案及解析】判断推理题。男士问有多少学生通过考试，女士说1016名学生参加考试，但一半没有通过，由此推断答案为C。

12. W: Would you tell me how I can get to the hospital?

M: Sure, let's me give you a hand.

Q: What's the man doing?

【答案及解析】判断推理题。女士问去医院的路，男士表示愿意帮忙。关键词是 give you a hand (给你提供帮助)，由此推断答案为D。

13. W: May I help you?

M: Yes, I'd like to call Jane's office, area code 502-2211, my name is Scott.

Q: Who is the man speaking to?

【答案及解析】判断推理题。女士问可以帮忙吗，男士回答，他要接 Jane 办公室的电话，同时报出号码和自己的名字，由此推断答案为D。

14. W: Is aunt Mary in? I've got something important to tell her.

M: Sorry, mother has gone shopping, she won't be back until lunch time.

Q: What is the relationship between the two speakers?

【答案及解析】判断推理题。女士要找 Mary 姑姑，男士说妈妈不在，由此推断两人关系为 cousin，故答案为D。

15. M: Sue isn't here yet, did you forget to invite her?

W: She was ready to come, but then changed her mind.

Q: Why isn't Mary present?

【答案及解析】细节考查题。男士问：Sue 不在，你没有邀请她吗？女士回答：她准备来，但又改变主意了。关键词是 change her mind，由此可知，答案为C。

16. W: Why are you just standing outside instead of going in?

M: I have tried all my keys in the lock, but it won't open.

Q: Why didn't the man go in?

【答案及解析】细节考查题。女士问男士为什么站在外面不进去，男士回答，他试了所有的钥匙，但是门打不开。关键是理解男士的回答。

17．M: You don't look over thirty.

   W: Really? In fact, I'm thirty-five.

   Q: How does the woman feel about the man's remark?

【答案及解析】判断推理题。男士说女士看上去不到 30 岁，女士回答：真的吗？事实上我已经 35 岁了。35 岁的女士听到别人说她不到 30，态度必然是 happy，由此推断答案为 A。

18．M: Did your sister like her new car?

   W: She thought it was too noisy, and something got wrong with the tires, but my father believed it was quite a good car.

   Q: What did the girl's father think of the new car?

【答案及解析】细节考查题。本题干扰的句子较多，问题是女孩父亲对汽车的态度，所以关键是女士回答的最后一句：我父亲认为这是一辆好车。由此得知，答案为 A。

Now you will hear two long conversations.

Conversation One

M: Susan, I could really use your help this weekend.

W: What is it, John? Another term paper?

M: No, no. This is easy compared to that. My cousin is coming on Thursday. She has an interview at the college and I promised my aunt I'd look after her. We are going to the game on Friday, but Saturday I'm on duty at the library all day and can't get out of it. Uh, I was wondering if you could show her around during the day and maybe we can all meet for dinner later.

W: Sure. I don't have any plans. What kind of things does she like to do?

M: Actually I haven't seen her for three years. She lives so far away. But this will be her first time on a college campus, she is still in high school. So she will probably enjoy anything on campus.

W: Well, there is a music festival in the auditorium. That's a possibility. Only I hope it doesn't snow. They are predicting 6~8 inches for the weekend. Everything will be closed down then.

M: Well, how about for the time being. I'll plan on dropping her off at your place on the way to work, around eleven. But if there is a blizzard, I'll give you a call and see if we can figure something else out.

W: Sounds good. Meantime I'll keep Saturday open. We can touch this Friday night when we have a better idea of the forecast.

M: I hope this works out. I feel kind of responsible. And I want her to have a good time. Anyway I really appreciate your help. I owe you one.

W: No problem. I'll talk to you tomorrow.

Questions 19 to 21 are based on the conversation you have just heard.

19．What will John do on Saturday?

【解析】事实细节题。男士在对话中说出了他星期六没时间陪他堂妹的缘由：but Saturday I'm on duty at the library all day and can't get out of it. (但是周六我得整天在图书馆值班，没时间出来。)

20．What does John ask Susan to do?

【解析】事实细节题。男士的堂妹周四要来，而他周五周六都没时间陪她，所以他对苏珊说: I was wondering

if you could show her around during the day. （希望苏珊陪他的堂妹逛逛，替他照料一下）。

21．What does John Say about his Cousin's interests?

【解析】推理判断题。在对话中女士问男士他堂妹喜欢什么，男士说：Actually I haven't seen her for three years. 这句话暗示他并不知道她喜欢什么。

Conversation Two

W: Wake up, Erik, time to rise and shine.

M: Ha, oh, hi, Jane. I must have fallen asleep while I was reading.

W. You and everyone else. It looks more like a campground than a library.

M: Well, the dorm's too noisy to study in, and I guess this place is too quiet.

W: Have you had any luck finding a topic for your paper?

M: No, Prof. Grant told us to write about anything in cultural anthropology. For once I wish she had not given us so much of a choice.

W: Well, why not write about the ancient civilizations of Mexico. You seem to be interested in that part of the world.

M: I am, but there is too much material to cover. I'll be writing forever, and Grant only wants five to seven pages.

W: So then limit it to one region of Mexico. Say the Uka town. You've been there and you said it has got lots of interesting relics.

M: That's not a bad idea. I brought many books and things back with the last summer that would be great resource material, now if I can only remember where I put them.

Questions 22 to 25 are based on the conversation you have just heard.

22．What was the man doing when the woman approached him?

【解析】事实细节题。女士对 Erik 说的第一句话：Wake up, Erik, time to rise and shine. (醒醒，天大亮了。)显然，此时男士正在睡觉。

23．Why has the woman come to talk to the man?

【解析】推理判断题。本题从整个对话的内容来分析，对话的场景地点是在图书馆，场景围绕着男士的论文选题展开。女士是来询问男士论文的进度的。

24．What seems to be the man's problem?

【解析】推理判断题。男士的问题是不知道写什么好，即不能确定他的论文选题，而 Prof. Grant 的选择太多了，让他不知道写什么了： Prof. Grant told us to write about anything in cultural anthropology. For once I wish she had not given us so much of a choice.

25．What is known about Prof. Grant?

【解析】推理判断题。关于 Prof. Grant 没明确的信息，但是从她给的论文选题范围可以推断出她从事的是人类学（anthropology）研究或教学人员，所以只有 C 正确。

Section B

Passage One

The first day of April ranks among the most joyous days in the juvenile calendar.

"It is a day when you hoax friends of yours with jokes like sending them to the shop for some pigeon's milk, or telling them to dig a hole because a dog has died, when they come back and ask where is the dead dog you say "April fool" and laugh at them .There are some when you just say "Your shoe lace is undone" or "Your belt is hanging" or "Go and fetch that plate off the table", and of course their shoe lace is tied up right, and their belt is not hanging , and there is no plate on the table, so you say "Ever been had, April fool." And parents, of course, are not exempt. "We have a lively time," says an 11-year-old Swansea girl, "as there are so many jokes to play such as sewing up the bottom of Daddy's trousers." And a 9-year-old Birmingham boy writes: "Last year I fooled father by gluing a penny to the floor and saying "Dad, you've dropped a penny on the floor." He couldn't get it off the ground because it was stuck firm, then I shouted "Yah, April fool ".

Questions 26 to 28 are based on the passage you have just heard.

【内容概要】本文主要叙述的是青少年在愚人节所搞的令人发笑的小动作，并举出详细的例子加以说明。

26. Which of the following is not true according to the passage?

【解析】文中谈到给鸽子买牛奶，给死了的狗挖个洞等，所没有谈到的是 D 项（在桌子上吃东西）。

27. Why did her father look for a penny?

【解析】文中说明，小孩子愚弄他的父亲，把一枚硬币粘在地板上，让他父亲认为是自己掉的，并让他父亲拣起来，所以选项为 C。

28. Why did she told her father that his penny dropped on the floor?

【解析】愚人节时，小孩子玩的小把戏都是愚弄别人，快乐自己。分析整篇文章，答案为 A。

Passage Two

Some months ago my friend bought a new refrigerator, but became worried when it did not work properly. The food did not keep well, the milk went sour quickly and the inside of the refrigerator had an unusual smell. My friend tried to do many different things to solve the problems. She checked every corner inside the fridge and wiped the refrigerator out with a wet cloth, all without success. This refrigerator would not work properly. Finally, my friend decided she had enough. She asked her son-in-law to find the paper with the shop guarantee on it. This guarantee said the shop promised to repair the refrigerator for free if it broke down in the first three months. The son-in-law first had a look at the refrigerator to see what the problem was. To the amusement of all those present and the embarrassment of my friend, she found that she had plugged her refrigerator into the electric power plug in the wall, but had forgotten to switch the power on.

Questions 29 to 31 are based on the passage you have just heard.

【内容概要】本文叙述的是我的朋友买了一台冰箱，但由于使用不当，未能发挥冰箱的效能，放进去的食物都坏了。结果经检查是未接通电源，而不是冰箱本身的问题。

29. Why does my friend become worried when she bought a new refrigerator?

【解析】文章前两段谈到，当我的朋友买回一台新的冰箱时，却不能正常使用，放进去的食物全部坏掉，他自然会很着急，故答案应为 B。

30. Which of the following is not true according to the passage?

【解析】文中谈到，放到冰箱中的牛奶很快变酸，冰箱内有异味，于是我的朋友检查了冰箱的每一个角落，并且将其彻底擦拭了一遍，故可得知 C 不正确。

31. Which of the following is written in the paper with the shop guarantee according to the passage?

【解析】文章第三段最后一句话谈到，商店售出冰箱三个月内可免费维修，得知答案为 C。

Passage Three

London taxi drivers know the capital like the back of their hands. No matter how small or indistinct the street is, the driver will be able to get you there without any trouble. The reason London taxi drivers are so efficient is that they all have gone through a very tough training period to get special taxi driving license. During this period, which can take two to four years, the would-be taxi driver has to learn the most direct route to every single road and to every important building in London. To achieve this, most learners go around the city on small motorbikes, practicing how to move to and from different points of the city. Learner taxi drivers are tested several times during the training period by government officers. The exams are terrible experience. The officers ask you "How do you get from Birmingham palace to the Tower of London? " and you have to take them there in the direct line. When you get to the tower, they won't say "well done". They will quickly move on to the next question. After five or six questions, they will just say "See you in two months' time" and then you know the exam is over. Learner drivers are not allowed to work and earn money as drivers. Therefore, many of them keep their previous jobs until they have obtained the license. The training can cost quite a lot, because learners have to pay for their own expenses on the tests and the medical exam.

Questions 32 to 35 are based on the passage you have just heard.

【内容概要】 本文是关于伦敦出租车司机的考核过程的介绍。伦敦的出租车司机要经过特殊的训练和严格的测验才能获得驾驶执照。

32．Why are London taxi drivers very efficient?

【解析】细节题。根据短文开头 The reason London taxi drivers are so efficient is that they all have gone through a very tough training period to get special taxi driving license.（……他们要经历很严格的训练期才能获得特别出租车驾驶证），答案选 B) Because they have received special training.

33．How long does the training period last?

【解析】细节题。根据文中 During this period, which can take two to four years 可判断答案为 A) Two years or more.

34．Why does the speaker think the driving test is a terrible experience?

【解析】细节题。Learner taxi drivers are tested several times during the training period by government officers. The exams are terrible experience.（学员司机在培训期间被政府官员反复的进行严格的测验）， 推出答案为 B) The learner has to go through several tough tests.

35．Why do learner drivers have to keep their present jobs?

【解析】细节题。Learner drivers are not allowed to work and earn money as drivers. Therefore, many of them keep their previous jobs until they have obtained the license.（学员司机不允许像司机一样工作和赚钱，因此他们都保留现有的工作直到拿到驾照。）推出答案是 C) They cannot earn money as taxi drivers yet.

Section C

In the north of Scotland, there is a deep, dark lake (36) surrounded by mountains. This is Loch Ness. Loch is the (37) Scottish word for "lake". Loch Ness is related to a big and strange creature that was said to live in the lake.(38) Although no one ever got a good look at it, the local people believed in this creature. They (39) thought it must be some kind of fish, since it lived in the lake.

Before the 1930's, few outsiders had heard of the creature. Then a road was built along Loch Ness and many (40) visitors came to see for themselves. Some believed they had caught sight of it. Many newspapers (41) printed stories about the creature.

These stories made the (42) creature famous. But some readers (43) considered it a joke. To them, this strange thing was an invented animal, something they might see in a movie.

(44) Accounts of the Loch Ness monster also sounded like jokes. Many people thought they had seen part of it. The parts added up to a very strange creature indeed. It was said to be 20 or 30 or 50 feet long. (45) The body was thick in the middle but thin at the tail. There was a long neck with a small head. Sometimes the back looked, like a boat turned upside down. At other times it had one, two or three humps like a camel. (46) The monster seemed shy. It never attacked people, and any noise caused it to disappear.

Part IV   Reading Comprehension (Reading in Depth)
Section A

【短文大意】我们认为的大气层大部分实际上是对流层，对流层就是离地球最近的大气层。大部分的气象变化就发生在这里，并且这里是唯一有足够氧气和热量供人类生存的大气层。这部分大气层在赤道上空厚度大约有10英里，而在地球的两极上空就只有5英里左右。

在对流层上面是平流层，如果你乘过国际航班，那么你就很有可能见过它。但在这种海拔高度上还有其他类型的"飞机"，比如被称为"急流"的气流，就穿梭在平流层。平流层中含有臭氧层，它能过滤对人类有危害的紫外线，否则这种对人体有害的射线直接射到地面上将使地球上的生命灭绝。在平流层上面是中间层，在中间层之上就是电离层，电离层最重要的作用就是让广播通讯信号从电离层发射到世界各地。

许多人认为大气层中大部分成分是氧气，因为这是我们所呼吸的气体。但实际上氧气仅占大气总体积的21%，而二氧化碳占的不足1%。占大气总体积超过3/4的是氮气，当地球还处于非常活跃的时候，氮气就从地球里面被释放出去。在我们人体中虽然存在大量的氮气，但我们不能从大气中直接摄取它。相反我们可以从我们所吃的植物中获取它。

47.【答案及解析】N) 文章从第一段到第二段按照高度的顺序（标志词是几个 above）介绍了大气层的组成部分。我们平时所说的大气层为对流层（troposphere），距离地球最近。

48.【答案及解析】G) 所填词为名词，由 our 所修饰，并根据 most 和 happens 判断该词不可数。在几个名词选项中，只有 weather 符合上下文，对流层是气象活动的大气层。

49.【答案及解析】L) 所填词与 to 构成动词不定式，而前面的 enough oxygen 和 warmth 对动词的含义进行了提示，氧气和热量都是生存（survive）所必需的。breathe 与 warmth 不相符，因而排除。

50.【答案及解析】A) 本题根据句内关系选择答案。上半句提到对流层有10英里厚，这里的厚度其实就是高度（height），而不是大小（size）。其中 half that + n.是一种常用的比较句型。

51.【答案及解析】E) 副词 otherwise 是假设一种与上文不同的情况，在这里引导了虚拟语气。如果没有臭氧层（ozone layer）的保护，有害的紫外线（ultraviolet rays）会对生命造成严重影响，句中的 make 暗含了一种因果关系，因此 extinct 符合语境。

52.【答案及解析】K) 无线传播信号遍布世界各地，因此选用 different 修饰 parts。

53.【答案及解析】B) 所填词作谓语，表明 we 与 oxygen 的关系，故选择 breathe。

54.【答案及解析】O) 空格中填入形容词作表语，说明地球（the planet）的情形。火山喷发、释放氮气的时候，地球处于比较活跃（active）的时期。

55.【答案及解析】D) 所填词在 get it 和 from 之间，为副词，根据后一句 Instead we get our nitrogen from plants 可以判断我们获取氮气的方式是间接的。因此选用 directly，与前面的 not 一起表示间接的意思。

56.【答案及解析】M) 所填词做谓语，表明 we 与 plants 的关系，故选择 eat 。

Section B

Passage one

If Europeans thought a drought was something that happened only in Africa, they know better now. After four years of below-normal rainfall (in some cases only 10 percent of the annual average), vast areas of France, Spain, Portugal, Belgium, Britain and Ireland are dry and barren. Water is so low in the canals of northern France that waterway traffic is forbidden except on weekends. Oyster（牡蛎）growers in Britain report a 30 percent drop in production because of the loss of fresh water in local rivers necessary for oyster breeding. In southeastern England, the rolling green hills of Kent have turned so brown that officials have been weighing plans to pipe in water from Wales. In Portugal, farmers in the southern Alentejo region have held prayer meetings for rain—so far, in vain. Governments in drought-plagued countries are taking drastic measures. Authorities in hard-hit areas of France have banned washing cars and watering lawns. In Britain, water will soon be metered, like gas and electricity. "The English have always taken water for granted," says Graham Warren, a spokesman of Britain's National Rivers Authority, "Now they're putting a price on it." Even a sudden end to the drought would not end the misery in some areas. It will take several years of unusually heavy winter rain, the experts say, just to bring existing water reserves up to their normal levels.

57. What does the author mean by saying "hey know better _____ now"?

A) They know more about the causes.

B) They have a better understanding of the drought in Africa.

C) They have realized that the drought in Europe is the most serious one.

细节 推断
D) They have realized that drought hit not only Africa but also Europe.

58. The drought in Europe failed to lead to the problem of_____.

细节 推断
A) below-normal rainfall

B) difficult navigation

C) a sharp drop in oyster harvest

D) bone-dry hills

59. The British government intends to_____.

A) forbid the car-washing service

B) increase the price of the water used

C) end the misery caused by the drought

同义 转述
D) put a price on water

60. Which of the following statements is TRUE according to the passage?

A) Germany is the only country free from the drought.

细节 推断
B) Water reserves are at their lowest level in years due to the drought.

C) The drought is more serious in Britain than in France.

D) Europe will not have heavy rain until several years later.

61. Which of the following is the most appropriate title for the passage?

A) Europe in Misery

B) Drought Attacks Europe （主旨大意）

C) Be Economical with Water

D) Europe, a Would-be Africa

【短文大意】如果欧洲人认为干旱是一件仅在非洲发生的事，他们现在该更清楚自己所面临的问题。在连续四年低于正常降雨之后(有时，仅仅是每年平均降雨量的10%)，法国、西班牙、葡萄牙、比利时、英国以及爱尔兰的大部分地区都很干旱和荒芜。法国北部的运河水位很低以至于河道交通除周末以外都被禁止。英国的牡蛎养殖者汇报，由于当地河流不能满足养殖牡蛎所必需的新鲜水源，造成牡蛎产量下降30%。在英国东南部的肯特郡，原本连绵起伏层峦叠翠的小山也变得如此苍黄，因而政府计划用管道从威尔士输送水源。在葡萄牙，南部的

104

Alentejo 地区的居民为求雨而举行祷告会，然而，到目前为止这些仍然是徒劳的。

那些被干旱困扰的国家，政府正在采取严厉的措施。法国灾情严重的地区的官方已经禁止洗车和灌溉草坪。在英国，像气和电一样，用水也要装水表了。"英国人总不把水当成一回事"，一个叫 Graham Warren 的英国国家水利权力机构发言人说，"现在他们要为用水而付钱了。"即使干旱灾害突然停止也不会结束一些地区的灾情。专家们声称仅仅把当前的水贮藏量提高到正常水平也得需要几年大规模的冬雨。

57.【题目译文】作者说"他们现在该更明白点了"的意思是_____？

【答案及解析】D) 第一段指出，如果欧洲人认为干旱仅是发生在非洲的事，他们现在该更明白点了。这一段紧接着的句子是从欧洲几个国家缺水这一现实说明了干旱不仅袭击非洲，欧洲也遭受了干旱的威胁。D) 与题意相符，故为正确答案。

58.【题目译文】欧洲的干旱未能导致的问题是_____。

【答案及解析】A) 根据第一段可以知道干旱带来的问题包括河道交通被禁止、牡蛎产量下降和山脉干旱苍黄，即选项B)、C) 和 D)。A) 低于正常降雨是导致干旱的原因，而不是干旱带来的问题。

59.【题目译文】英国政府计划_____。

【答案及解析】D) 由第二段第三句可知，在英国，像气和电一样，用水也要装水表了。下文 Warren 的话对此句进行了解释："英国人总不把水当成一回事，现在他们也对水定价交费了。"D) 与题意相符，故为答案。

60.【题目译文】根据短文所述，下面哪项陈述是正确的？

【答案及解析】B) 见文章最后两句话：Even a sudden end to the drought would not end the misery in some areas. It will take several years of unusually heavy winter rain, the experts say, just to bring existing water reserves up to their normal levels. 即使干旱灾害突然停止也不会结束一些地区的灾情。专家们声称：仅仅把当前的水贮藏量提高到正常水平也得需要几年大规模的冬雨。由此可以推断，几年来，水位已降至最低，故 B)为正确答案。

61.【题目译文】下面哪个是这篇短文最合适的标题？

【答案及解析】B) 主旨题。文章首尾呼应，文章第一句及最后两句概括总结了本文的主旨，即干旱正威胁着欧洲。

Passage Two

Real policeman hardly recognize any resemblance between their lives and what they see on TV—if they ever get home in time. There are similarities, of course, but the cops don't think much of them.

The first difference is that a policeman's real life revolves round the law. Most of his training is in criminal law. He has to know exactly what actions are crimes and what evidence can be used to prove them in court. He has to know nearly as much law as a professional lawyer, and what is more, he has to apply it on his feet, in the dark and rain, running down an alley after someone he wants to talk to.

Little of his time is spent in chatting to

62. It is essential for a policeman to be trained in criminal law_____.

A) so that he can catch criminal everywhere

B) because many of the criminals he has to catch are dangerous

C) so that he can justify his arrests in court

D) because he has to know nearly as much about law as a professional lawyer

63. The everyday life of a policeman or detective is_____.

A) exciting and glamorous

B) dangerous and venturous

C) devoted mostly to routine matters

D) wasted on unimportant matters

64. When murders and terrorists attacks occur, the police_____.

105

charming ladies or in dramatic confrontations with desperate criminals. He will spend most of his working life typing millions of words on thousands of forms about hundreds of sad, unimportant people who are guilty—or not—of stupid, petty crimes.

Most television crime drama is about finding the criminal: as soon as he's arrested, the story is over. In real life, finding criminals is seldom much of a problem, Except in very serious cases like murders and terrorist attacks, little effort is spent on searching.

Having made an arrest, a detective really starts to work. He has to prove his case in court and to do that he often has to gather a lot of different evidence. So, as well as being overworked, a detective has to be out at all hours of the day and night interviewing his witnesses and persuading them, usually against their own best interests, to help him.

A) prefer to wait for the criminal to give himself away

细节推断 B) spend a lot of effort on trying to track down the criminals

C) try to make a quick arrest in order to keep up their reputation

D) usually fail to produce results

65. Which of the following is TRUE according to the passage?

A) Generally the detective's work is nothing but to arrest criminals.

B) Policemen feel that the image of their lives shown on TV is not accurate.

C) People are usually willing to give evidence.

D) Policemen and detectives spend little time at the typewriter.

66. Which of the following is the most suitable title for this passage?

A) Real Life of a Detective

B) Detective's life—Fact and Fantasy

C) The Reality of a Detective

D) Policemen and Detective (主旨题)

【短文大意】
真正的警察都很难发现自己的生活和电视里所看到的警察生活之间有任何相同之处——如果他们能及时赶回家看电视的话。当然，也有一些相似的地方，但警察们往往认为不值一提。

两种生活的第一个不同之处就是，现实中的警察生活总是围绕着法律展开的。他的大部分训练均涉及刑法。他必须准确知道什么样的行为是犯罪，什么样的证据可以用在法庭上。他应该像职业律师那样熟悉法律，而且他必须外出办案，在实践中运用这些法律知识，不管是披星戴月，风餐露宿，还是奔走于街头巷尾去追踪他需要谈话的人。

警察很少有时间和那些迷人的女子聊天，也不大会富有戏剧性地与亡命之徒正面交锋。他的大多数工作时间都用来在成千上万份的表格上打出数百万文字的材料，报告那些可怜的小人物们所犯的（或者没犯的）愚蠢、轻微的罪行。

大部分有关犯罪的电视剧都侧重于寻找罪犯的过程：罪犯一旦被捕，故事就结束了。在现实生活中，寻找罪犯不是大问题。警方一般不在搜寻罪犯上花费太大力气，除非是谋杀和恐怖活动这样的要案。

一旦罪犯逮捕归案，侦探就真正开始工作了。他必须在法庭上查验案件，为此往往需要搜集大量不同的证据。因此，侦探不但要超负荷工作，还要夜以继日地四处走访证人，说服他们，违背自身的利益来帮助他。

62.【题目译文】警察很有必要在刑法方面接受培训_____。

【答案及解析】C) 第二段第二句和第三句指出，警察所受的大多数训练是刑法方面的，他们得确切知道哪些行为是犯罪，哪些证据可用在法庭上作证。C)与题意相符，故为正确答案。

63.【题目译文】警察或侦探的现实生活是_____。

【答案及解析】C) 第三段第二句指出，警察的大部分工作时间都用来在成千上万的表格上打字，用数以百万计的文字记录成百上千宗痛苦的、不重要的人在一些愚蠢的小案中是否有罪。由此可见，警察大量的时间都花在例行公事上，故 C)为正确答案。

64.【题目译文】当谋杀和恐怖活动这样的要案发生时，警察_____。

【答案及解析】B) 由第四段可知，大多数犯罪电视剧是关于找到罪犯的：一旦罪犯被捕，故事也就结束了。而在现实生活中，寻找罪犯很少成为问题。警察很少花力气寻找罪犯，除了在非常严重的案子上，如谋杀和恐怖分子的袭击。由此可见，只有在非常重大的案子，如谋杀和恐怖袭击上警察才会花大力气寻找罪犯，B)与题意相符，故为正确答案。

65.【题目译文】根据文章的内容，以下哪个是正确的选项？

【答案及解析】B) 文章第一句指出，真正的警察生活和他们在电视上看到的电视剧中的警察生活之间没有共同点。由此可见，警察认为电视上演的和他们的真实生活并不一致，故 B)为答案。

66.【题目译文】下面哪个选项最适合作本文的标题？

【答案及解析】B) 主旨题。本文主要阐述真正的警察生活与电视剧中警察生活的不同，生活是现实，而电视剧是幻想虚构的（fact and fantasy），B)很好地概括了这一主旨，故为正确答案。

Part V   Error Correction

【短文大意】庆祝母亲节的最好方式之一就是让母亲放假，让她放松，让其他的家庭成员工作。

许多家庭的母亲节早餐是在床上进行的。 通常父亲和孩子让妈妈睡觉，他们走到厨房准备妈妈最喜欢的早餐。母亲节的早餐可以包括母亲喜欢的任何一种食品。食物做好之后，他们在盘子上摆好所有的东西，不要忘记在花瓶里放一束鲜花，春天了，孩子们会到园子外面采郁金香或水仙花。

当所有东西都准备好后，他们小心地把盘子和妈妈最喜欢的东西拿到卧室。孩子们把准备的卡片和小礼物放在盘子上，之后送到妈妈床边。

许多家庭会在家为母亲准备母亲节的特殊晚餐，或者请母亲到餐馆吃饭，这是让母亲好好放松并且让她看到她拥有一个多么美好的家庭的一天！

67. ease→easy 考查固定短语 take it easy。

68. began→begin 考查时态。

69. later→late 考查形似词辨析。late "晚的"； later "稍候"。

70. consist ^ anything → of 考查固定搭配 consist of。

71. nice→nicely 考查词性。

72. are→is 考查主谓一致。

73. present→presents 考查单复数，句中的 present 作名词使用，意为"礼物"。

74. present→presented 考查被动语态，句子中的 present 作动词使用。

75. and→or 考查篇章理解。根据该句含义可判断"在家为母亲准备母亲节的特殊晚餐"和"请母亲到餐馆吃饭"是一种选择关系，而非并列关系。

76. how→ what 考查习惯用法。

Part VI   Translation

77. _____(平均来看), men smoke more cigarettes than women.

【题目译文】平均来看，男性比女性吸的香烟更多。

【答案及解析】On average：副词，平均。Eg. This car runs 15 kilometers per liter on average. 这辆车平均每公

升（汽油）跑 15 千米。

78. Your losses in trade this year are nothing_____(与我的相比).

【题目译文】你今年在贸易上的损失与我的相比，不足为奇。

【答案及解析】compared with mine / in comparison with mine 注意 compared with 和 in comparison with 的用法，表示"与…相比"。

79. Not only _____(他向我收费过高), but he didn't do a good repair job either.

【题目译文】他不仅向我收费过高，而且他的修理工作做得也不好。

【答案及解析】did he charge me too much / did he overcharge me Not only..., but also...：不仅……而且。还要注意倒装的使用方法。

80. Mary has to work with people of all ages _____(各行各业).

【题目译文】玛丽不得不和各行各业各个年龄段的人打交道。

【答案及解析】in all walks of life. "in"：介词，指在某方面。

81. The children _____(退学) and went to work .

【题目译文】孩子们退学去打工了。

【答案及解析】dropped out of 注意 dropped out of 有"退学，辍学"的意思。

大学英语 4 级考试新题型模拟试卷七答案及详解

参考答案

Part I　Writing

Desk Culture

Sitting in a classroom at college, you may often catch the sight of something written on the desk, which is an index of some college students' psychology and becomes part of the campus culture. This is why it is called "desk culture".

Desk culture has its substantial content. On most occasions they are definitions, formulas, or English expressions kept either for memory, for convenience because they have no paper at hand, or because they have them for a special "service"—to provide information for cheating on exams. Sometimes there are also some ragged verses, drawings or even dirty words created by some naughty or sentimental students, made for joking or expressing their own feelings. Occasionally there are some mottos for encouragement, or some senseless marks habitually made to kill time when they are bored or absent-minded in class. In a word, all these things contribute to the flourishing desk culture.

As far as I am concerned, desk culture indirectly indicates college students' mental orientation. It is, for whatever reasons, at least a destruction of the perfection of the desks. So we should try to avoid this kind of destruction, or find a better way to express our feelings and colleges should put more emphasis on quality education and spiritual civilization.

Part II　Reading Comprehension (Skimming and Scanning)

1. Y　　2. NG　　3. Y　　4. NG　　5. N　　6. Y　　7. N

8. 830,000　　9. volcanic eruptions　　10. doorways or even bathtubs

Part III　Listening Comprehension

Section A

11. C　12. D　13. C　14. B　15. D　16. A　17. B　18. C　19. B　20. C

21. A　22. B　23. C　24. A　25. C

Section B

26.D　27.A　28.C　29.D　30.C　31.D　32.A　33.D　34.D　35.D

Section C

36.birthplace　　37. queue　　38. entirely　　39. emigrated

40. interest　　41. members　　42. fortunate　　43. possible

44.I had taken Shakespeare for granted　　45.Respect was called for

46.An inclusive ticket enabled holders to see them all

Part IV　Reading Comprehension (Reading in Depth)

Section A

47. F: establish　　48. E: fierce　　49. N: point　　50. I: express　　51. H: going in for

52. L: communicate　53. J: principles　54. K: Agency　55. D: attention　56. B: sums

Section B

57. C　58. A　59. C　60. B　61. D　62. D　63. D　64. C　65. B　66. D

Part V　Cloze

67.A　68. B　69. B　70.D　71. B　72. C　73.A　74.D　75. B　76.A

77.C 78. D 79.A 80. B 81. D 82. B 83.C 84. A 85. D 86. D

Part VI　　Translation

87. whether(it is) heated or not

88. follow your lead

89. give of yourself

90. look up to

91. take the place of

答案解析及录音原文

Part I　　Writing

Desk Culture

Sitting in a classroom at college, you may often catch the sight of something written on the desk, which is an index of some college students' psychology and becomes part of the campus culture. This is why it is called "desk culture".

Desk culture has its substantial content. On most occasions they are definitions, formulas, or English expressions kept either for memory, for convenience because they have no paper at hand, or because they have them for a special "service"—to provide information for cheating on exams. Sometimes there are also some ragged verses, drawings or even dirty words created by some naughty or sentimental students, made for joking or expressing their own feelings. Occasionally there are some mottos for encouragement, or some senseless marks habitually made to kill time when they are bored or absent-minded in class. In a word, all these things contribute to the flourishing desk culture.

As far as I am concerned, desk culture indirectly indicates college students' mental orientation. It is, for whatever reasons, at least a destruction of the perfection of the desks. So we should try to avoid this kind of destruction, or find a better way to express our feelings and colleges should put more emphasis on quality education and spiritual civilization.

【评论】本文第一段引出文章话题，即什么是课桌文化，然后在第二段列举了其内容及产生的原因，最后一段表明了自己的看法。本文层次清晰，语言生动，句式结构多变，读起来朗朗上口。

【常用句型】

第一段：点明问题

This is why it is called…

第二段：阐述观点

In a word, all these things contribute to…

第三段：点明自己的观点

As far as I am concerned …

Part II　　Reading Comprehension (Skimming and Scanning)

【文章及答案解析】本文围绕有关地震的问题展开，全文开篇讲述了地震的成因，然后对地震的灾害及如何，预测及应急。

What causes earthquakes? The earth is formed of layers. The surface of the earth, about 100 kilometers thick, is made of large pieces. When they move against each other, an earthquake happens. A large movement causes a violent earthquake, but a small movement causes a mild one.

细节
推理

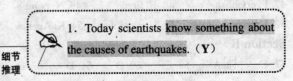

1. Today scientists know something about the causes of earthquakes. （**Y**）

110

Earthquakes last only a few seconds. The rolling movements are called seismic waves. The seismic waves start in one place, called the epicenter, and roll outward. A seismic wave travels around the earth in about twenty minutes. Usually, an earthquake is strong enough to cause damage only near its epicenter.

However, epicenters at the bottom of the ocean create huge sea waves as tall as 15 meters. These wavescross the ocean in several hours. Rushing toward land, they destroy small islands and ships path. When they hit land, they flood coastal areas far from the epicenter of the earthquake. In 1868, a wave reached 4.5 kilometers inland in Peru. In 1896, a wave in Japan killed 27,000 people.

After an earthquake happens, people can die from lack of food, water, and medical supplies. The amount of destruction caused by an earthquake depends on where it happens, what time it happens, and how strong it is. It also depends on types of buildings, soil conditions, and population. Of the 6,000 earthquakes in the world each year, only about fifteen cause great damage and many deaths.

In 1556, an earthquake in northern China killed 830,000 people—the most in history. There was no way to measure its strength. In 1935, scientist started using the Richter Scale to measure seismic waves. A seriously destructive earthquake measures 6.5 or higher on the Richter Scale.

How can scientists predict earthquakes? Earthquakes are not just scattered anywhere but happen in certain areas, places where pieces of the Earth's surface meet. This pattern causes them to shake the same places many times. For example, earthquakes often occur on the west coast of North and South American, around the Mediterranean Sea, and along the Pacific coast of Asia.

Another way to predict earthquakes is to look for changes in the earth's surface, like a sudden drop of water level in the ground. Some people say animals can predict earthquakes. Before earthquakes, people have

【文章大意】

为什么会发生地震？地球由层面构成。形成地壳的巨大板块厚度约有 100 千米。板块间的相互运动，就形成了地震。大规模的板块运动会引发严重地震，而轻微的振动则影响不大。

地震的持续时间只有几秒钟，地壳的起伏运动叫震波；震波从某一地点即震源处开始，以波动的形式向外传播，20 分钟内传遍整个地球。通常地震造成的破坏仅限于震源附近。

细节
推理

2．More people are killed by huge sea waves than by buildings falling. (NG)

然而，位于海底的震源会引发高达 15 米的巨浪。巨浪数小时内波及整个地面。在向陆地涌入时，一路毁坏途径的岛屿和船只。一旦登陆，会使远离震源的沿海地区发生洪涝灾害。1868 年，海浪对秘鲁内陆的纵深冲击达 4.5 公里。1896 年，发生在日本的海浪造成 27,000 人丧生。

同义
转述

3．The vast majority of the world's earthquakes are mild. (Y)

8. The death toll of the earthquake in northern China — the most in history — reached＿＿＿.(830,000)

地震发生后，人们会由于缺少食物、饮用水和药品而死亡。其损失的程度取决于地震发生的地点、时间和强度，也取决于建筑物的类型、土质和人口数量。每年世界约发生 6,000 次地震，造成巨大损失和人口伤亡的地震只有 15 次左右。

1556 年，华北地震中死亡人数高达 830,000 人——是历史之最。那时还没有办法测量地震的强度。到了 1935 年，科学家们开始用里氏震级来测量震波。严重破坏性地震的指数为 6.5 级或更高。

科学家是如何预测地震的？地震不是分散各处而是集中发生在板块碰撞的某一地区或地点。这种模式造成它们在同一位置多次振动。如北美洲和南美洲的西海岸、地中海沿岸及亚洲的太平洋沿岸都是经常发

seen chickens sitting in trees, fish jumping out of the water, snakes leaving their holes, and other animals acting strangely.

On February 4, 1975, scientists predicted an earthquake in northeastern China and told people in the earthquake zone to leave the cities. More than a million people moved into the surrounding countryside, into safe, open fields away from buildings. That afternoon, the ground rolled and shook beneath the people's feet. In seconds, 90 percent of the buildings in the city of Haicheng were destroyed. The decision to tell the people to leave the cities saved 10,000 lives.

However, more than a year later, on July 28, 1976, the scientists were not so lucky. East of Beijing, Chinese scientists were discussing a possible earthquake. During their meeting, the worst earthquake in modern times hit. Estimates of deaths ranged from 250,000 to 695,000. The earthquake measured 7.9 on the Richter Scale.

Earthquakes often come together with volcanic eruptions. In late 1984, strong earthquakes began shaking the Nevado del Ruiz volcano in Colombia every day. On November 14, 1985, it erupted. A nearby river became a sea of mud that buried four towns. This disaster killed more than 2,100 people.

Mexico City has frequent earthquakes. An earthquake there on September 19, 1985, measured 8.1 on the Richter Scale and killed 7,000 people. Most victims died when buildings fell on them.

San Francisco, California, also has frequent earthquakes. However, newer buildings there are built to be safe in earthquakes. Therefore, when an earthquake measuring 7.1 on the Richter Scale hits northern California on October 17, 1989, only 67 people were killed. The earthquake hit in the afternoon, when thousands of people were driving home from work. Freeways and bridges broke and fell. Buried under the layers of the Oakland Freeway, people were crushed in their flattened cars. Explosions sounded like thunder as older buildings seemed to burst apart along with the freeways. As the electric power lines broke from the

生地震的区域。

另一种预测地震的方法是查看地表的变化，如地面水位的突然下降。有些人说动物可以预报地震。地震发生前，可以发现鸡飞上树，鱼跃出水面，蛇离开洞穴，其他动物行为异常。

1975年2月4日，科学家预测出中国东北部有地震并通知住在震区的人们撤离城市。一百多万人撤离到乡村附近或远离建筑物的安全开阔地。下午，人们脚下的地面开始颤抖摇晃。几秒钟内，海城90%的建筑全部被毁坏。通知人们撤离的决定拯救了10,000人的生命。

然而，一年多后，1976年7月28日，科学家们则没有那么好运。中国的科学家正在北京东部讨论地震发生的可能性，就发生了近代最为严重的地震。估计死亡人数为250,000～695,000人，里氏震级7.9级。地震常伴随火山喷发。1984年末，强烈的地震开始每天震动位于哥伦比亚的鲁伊斯火山。1985年11月14日火山爆发。附近的一条河流变成泥海，吞噬了四个城镇，灾难造成2100多人死亡。

9. Earthquakes often come together with _____. (**volcanic eruptions**)

墨西哥城经常发生地震。1985年9月19日发生的8.1级地震，有7000人丧生。其中大部分受害者死于在建筑物的倒塌。

加利福尼亚州的旧金山市也经常发生地震，但新的建筑物在地震中却很安全。因此，1989年10月17日，发生在加利福尼亚洲北部的里氏7.1级地震中，只有67人死亡。地震发生在下午，当时很多人正驱车回家。高速公路和大桥破损后塌陷。人们被压在扁平的汽车里，埋葬在奥克兰高速公路层下。老旧的建筑物似乎随着高速公路一起爆裂，声音听起来像在打雷。由于倒塌的桥梁和建筑物中的电线已破损，天空在大片黑色尘云的笼罩下似乎带着闪电。破裂水管中流出的水冲向街道，混着泄漏的煤气引起了更大的爆炸。

细节辨认

4. An earthquake in 1989 destroyed the city of Oakland. (**NG**)

falling bridges and buildings, the sky, covered with huge clouds of black dust, appeared to be filled with lightning. Water rushed into the streets from broken pipes and mixed with gas from broken gas lines, causing more explosions.

Emergency workers had to cope with medical problems. Everyone worked together to save survivors and comfort victims. The next day, the disaster sites looked terrible. Victims couldn't find their houses, their cars, or even their streets. Boats were destroyed, and debris covered the surface of the sea. There was no water, no electricity, no telephone only the smell of garbage floating in melted ice in refrigerators open to the sun. Losses and property damage from the earthquake amounted to millions of dollars.

Seismology is the study of earthquakes, and a seismologist is a scientist who observes earthquakes. Seismologists have given us valuable knowledge about earthquakes. Their equipment measures the smallest vibration on the surface of the earth. They are trying to find ways to use knowledge about earthquakes to savelives and to help solve the world's energy shortage. The earth's natural activity underground creates energy in the form of heat. Geothermal means earth heat. This geothermal energy could be useful. However, if we take natural hot water out of the earth in earthquake zones, we might cause earthquakes.

People live in earthquake zones because of natural beauty, productive soil, and large existing centers of population. However, people who live there should expect earthquakes. They should be prepared to protect their lives and property. They must build safer buildings and roads. Hospitals and electric power stations must be built as far as possible from probable earthquake sites. When an earthquake starts, people must run to open ground or stay in protected areas like doorways or even bathtubs.

If seismologists could predict earthquakes, we could save about 20, 000 human lives each year. Human can control many things about nature, but we can not control earthquakes.

急救人员必须解决医疗问题。大家共同努力抢救幸存者和安慰受害者。第二天，灾区看起来很可怕。受害者无法找到他们的房屋、汽车，甚至是居住的街道。船只遭到破坏，碎片漂浮在海面上。没有水、没有电，也没有电话，只有太阳照射下，漂浮在冰箱里融冰中垃圾的味道。地震造成的财产损失达数百万美元。

地震学是研究地震的学科，而地震学家是观察地震的学者。地震学家给我们带来了地震方面的宝贵知识。他们的仪器可以测量地表上最微小的颤动。他们正试图利用地震知识挽救生命和帮助解决全球能源短缺问题。地球在地表下的自然运动会以热的形式产生能量。地热就是地球热量，地热能用处广泛，但假使我们在地震带抽取天然热水，则可能引发地震。

细节辨认 5. Seismologists can measure the size of sea waves. (**N**)

由于地震带有美丽的自然风光、肥沃的土壤和大量已存在居民点，有很多人在那里生活。但居住在那里的人们应该学会预测地震，做好保护生命财产的准备，修筑更好安全的建筑和公路。医院和电厂必须修建在远离可能发生地震的地方。当地震发生时，人们可以跑到开阔地或停留在安全区，如门口或是浴缸。如果地震学家能够预测地震，我们每年就能挽救20,000 人的生命。人类能够控制关于自然的很多事情，但我们无法控制地震。

7. The passage gives a general description of the earthquakes' destruction. (**N**)

6. Removing water from underground may cause earthquakes. (**Y**)

同义转述 10. When an earthquake occurs, people must run to open ground or stay in safe areas like_____. (**doorways or even bathtubs**)

113

1.【题意】科学家对地震的成因有所了解。

【解析】本文将围绕地震的问题展开,首句以设问开头(What causes earthquake?),接着说明了形成地震的原因,地表不同板块的运动会引起地震,从中可以看出,科学家对地震的成因有所了解,即 know something about the causes of earthquakes,故选 YES。

2.【题意】同建筑物倒塌相比,海浪会造成更多的人死亡。

【解析】题干将海浪与建筑倒塌造成人员死亡相比较,因此在文章中应找出 huge sea waves 以及 building falling 的信息。第三段列举了海浪的影响,第十一段提到了 most victims died when building fell,但全文没有将二者做比较,无法得出哪个造成的死亡人数更多,故此题选 NG。

3.【题意】世界上大多数的地震都是轻微的。

【解析】在第三段的句末指出世界每年有 6,000 次地震,其中 15 次左右将会造成巨大的损失和死亡。因此,大多数地震都是轻微的。如果能对地理知识有所了解,就不会造成对题干的误解,故此题选 YES。

4.【题意】1989 年的地震毁灭了奥克兰市。

【解析】文章倒数第五段提到 buried under the layers of the Oakland Freeway,但没有信息表明 Oakland 是被地震摧毁的,所以不要将 Oakland Freeway 与 Oakland 相混淆,故此题选 NG。

5.【题意】地震学家可以测量海浪的强度。

【解析】题干涉及到 seismologists 的定义,倒数第三段句首提到了 seismologists 的定义,接着说明 seismologists 是研究地震的,故此题选 NO。

6.【题意】地下水的抽取能够引起地震。

【解析】倒数第三段最后一句由 however 引起,指出 take natural hot water out of earth in earthquake zones,可能引起 earthquake,题干用 removing water from underground 涵盖了原文的信息,may 与 might 同义,故该题选 YES。

7.【题意】文章概括地描述了地震的破坏性。

【解析】本题涉及到文章的主旨,全文开篇讲述了地震的成因,然后对地震的灾害及如何预测、怎么应急都做出了阐述,题干对主题的概括过于片面,the earthquake's destruction 只是文章的部分内容。

8.【题意】历史上最强烈的一次地震发生在华北造成的人员伤亡达 83 万。

【解析】题干提到历史上造成死亡人数最多、发生在华北的一次地震,在原文中可以找到 the most in history 等相关的信息,可以确定死亡数字为 830,000。题干用 the death toll reached 对原文中的 killed 做了同义转述,故此题应填 830,000。

9.【题意】地震的发生常伴随火山爆发。

【解析】在文中可以找到与题干完全相同的部分 Earthquakes often come together with,故该空应填 volcanic eruptions。

10.【题意】当地震发生时,人们可以跑到开阔地或呆在安全区如门口甚至浴缸里。

【解析】文章倒数第二段最后一句话指出 When an earthquake starts, people must run to open ground or stay in protected areas like doorways or even bathtubs。其中,starts 与题干的 occurs 同义,protected 与 safe 同义,故该题应填 doorways or even bathtubs。

Part Ⅲ   Listening   Comprehension

Section A

11. W: Are you busy today?

M: Yes. I have four classes in the morning and three in the afternoon.

Q: How many classes does the man have?

【答案及解析】本题属于推理题。女士说："你忙吗"，男士回答："是的，上午有四节课，下午有三节课。" 由此得出一共七节课。

12. M: What a delicious cake! Did you bake it?

W: No, I had the bakery do it.

Q: What does the woman mean?

【答案及解析】本题属于推理题。男士说："多可口的蛋糕啊! 是你烤的吗?"女士回答："不，不是，是我让面包店做的。"由 "bakery 面包店"一词可推出答案是 D）面包是烤面包店烤的。A）她自己烤的，B）她从商店买的。C）她想让那个男人给她烤。 A、B、C 均属错误选项。

13. M: Would you like some more potatoes?

W: I'm sorry I can't manage more. Thank you.

Q: What does the man ask the woman to do?

【答案及解析】本题属于推理题。男士说："你想再吃些土豆吗？"女士回答："我吃不下了"。由此可推出答案是 C）吃土豆。错误项 A) 买土豆。B）递给他土豆。D）帮他做土豆。

14. W: Jim looks nice in that new shirt, doesn't he?

M: I still wish he'd dress in that old shirt.

Q: What does the man mean?

【答案及解析】本题属于推理题。女士说："吉姆穿新衬衣很漂亮，不是吗？"男士说："我仍希望他穿旧的。" 由此可推出答案是 B）他穿新衬衣不好看。错误项 A）穿什么都好看。C）穿旧的不好看。D) 他让吉姆借给他衬衣。

15. M: Has the rain stopped?

W: Stopped? Look at my clothes, they are soaked.

Q: What does the woman mean?

【答案及解析】本题属于推理题。男士说："雨停了吗？"女士说："停? 看我的衣服都浸透了"。soak: 浸透。由此可推出答案是 D）雨下得更大了。错误项 A）雨停了。B）她想要浸透她的衣服。C）她在找衣服。

16. M: Oh, I forgot to bring my notebook.

W: Don't worry. You can borrow some paper from me.

Q: What does the woman mean?

【答案及解析】本题属于推理题。男士说："我忘记带笔记本了。"女士说："我可以借你几张纸"由此可推出答案是 A） 她能给他一些东西写字。错误项 B）她不知道谁拿走了他的笔记本。C）她想给他借一些纸。D）他可以从她那借个笔记本。

17. M: We can all go swimming after class tomorrow.

W: If it's a nice day, of course.

Q: What does the woman mean?

【答案及解析】本题属于推理题。男士说："明天下课后我们可以去游泳。"女士说："如果是晴天，当然可以。" 由此可推出答案是 B）依天气而定。错误项 A）她想改天去。C）是去上课的好天气。D）他们中有一些人能去游泳。

18. W: The 10:30 train is late again.

115

M: No surprise in such a bad weather.

Q: What does the man mean?

【答案及解析】本题属于推理题。女士说："10:30 的火车又晚点了。"男士说："在这样的坏天气里也不是奇怪的事情。"由此可推出答案是 C）火车晚点是因为天气不好，错误项 A）天气不好，他很惊讶。B）他认为火车不能晚点。D）没人对火车晚点奇怪。

Now you will hear two long conversations.

Conversation One

W: I'm thinking about transferring out of State College into another school in the spring.

M: After a year and a half? How come? I thought you liked it here.

W: I do. But our commercial art department doesn't give Bachelor's degrees, only associate. I want a Bachelor's.

M: So where do you want to go?

W: I wouldn't mind going to West water University. It has an excellent reputation for commercial art, but I have a feeling it's very selective.

M: But you've gotten good grades in the three semesters you've been in the state college, haven't you?

W: Yeah, mostly As and Bs.

M: So what are you worded about, just ask your professor to write letters of recommendation for you, and you'll be set.

W: Don't mention it.

Questions 19 to 21 are based on the conversation you have just heard.

19. What is the woman planning to do?

【解析】事实细节题。女士的第一句话就说明了她的打算：I'm thinking about transferring out of State College into another school in the spring.（我想在春季从州立学院转到另一所学校）。

20. How long has the woman been studying in State College?

【解析】事实细节题。听到女士要转学的消息，男士很吃惊：After a year and a half?（在学了一年半之后转学?）

21. What does the man suggest the woman do?

【解析】事实细节题。男士的最后一句话为女士转学提了建议：just ask your professor to write letters of recommendation for you（请你的教授给你写推荐信）。

Conversation Two

M: Uh... excuse me, Ms. Sherwin, but I was wondering if I could speak to you for a few minutes.

W: Well, I'm rather busy at the moment, Jerry. Is it urgent?

M: Uh, yes, I... I'm afraid it is. It's a personal matter.

W: Oh, well, then, we'd better discuss it now. Sit down.

M: Thank you. Uh... you see, it's about my wife. She... uh... well.., she...

W: Yes, go on, Jerry. I'm listening.

M: She's ill and has to go to hospital tomorrow. But we have a young baby, you know.

W: Yes, I know that, Jerry. You must be rather worried. Is it anything serious? Your wife's illness, I mean?

M: The doctors say it's just a minor operation. But it has to be done as soon as possible. And... well.., the problem is my daughter. The baby. That's the problem.

W: In what way, Jerry? I'm not quite sure if I understand.

M: Well, as I said, my wife will be in hospital for several days, so there's nobody to look after her.

W: You mean, nobody to look after your daughter, is it that?

M: Yes, exactly. Both our parents live rather far away, and.., and that's why I'd like to have a few days off from tomorrow.

W: I see. I think I understand now. You need a few days off to look after your daughter while your wife is in hospital.

M: Yes, yes. That's it. I'm not explaining this very well.

W: No, no. On the contrary, I just want to be sure I understand completely. That's all.

M: Will... will that be all right?

W: Yes, I'm sure it will, Jerry. All I want to do now is to make sure that there's someone to cover for you while you're away. Uh... how long did you say you'll need?

M: Just a few days. She... my wife, I mean.., should be out of hospital by next Thursday, so I can be back on Friday.

W: Well, perhaps you'd better stay at home on Friday, as well. Just to give your wife a few extra days to rest after the operation.

M: That's very kind of you, Ms. Sherwin.

W: Don't mention it.

Questions 22 to 25 are based on the conversation you have just heard.

22. What is the possible relationship between the two speakers?

【解析】推理判断题。答题的关键是判断对话中的场景。情节是这样的：Jerry 的妻子因病需要住院几天，女儿在家没人照料（nobody to look after her），所以他需要几天假在家照顾他的女儿。由此判断二人是老板与雇员的关系。（雇员在向老板请假。）

23. Where does the dialogue most probably take place?

【解析】推理判断题。根据上题的分析，地点是在老板的办公室。

24. Why does Jerry want to see Ms. Sherwin?

【解析】事实细节题。根据上题的分析，场景是雇员在向老板请假，所以 Jerry 的目的是请假。

25. When will Jerry be back to work?

【解析】推理判断题。Jerry 原本打算下星期五回来上班（so I can be back on Friday），但老板又宽限他几天（perhaps you'd better stay at home on Friday, as well. Just to give your wife a few extra days to rest after the operation.）老板说周五不必回来上班，好让他的妻子手术后多休息几天。

Section B

Passage One

Florida International University has opened what it says is the first computer art museum in the United States. You don't have to visit the university to see the art. You just need a computer linked to a telephone. You call the telephone number of a university computer and connect your own computer to it. All of the art is stored in the school computer. It is computer art, created electronically by artist on their own computers. In only a few minutes, your computer can receive and copy all the pictures and drawings. Robert Shasta is director of the new computer museum. He says he started the museum because computer artists had no place to show their work and he just want to help them to some extent.

A computer artist could only record his pictures electronically and send the records, or floppy discs, to others to see on their computers. He could also put his pictures on paper. But to print good pictures on paper, the computer artist needed an expensive laser printer.

Questions 26 to 28 are based on the passage you have just heard.

【内容概要】佛罗里达的一所大学创办了美国第一家电脑艺术博物馆，你只需拨通大学的一台电脑的号码，把自己的电脑和它连起来，就可以欣赏到艺术家们在电脑上完成的杰作了。如果需要，还可以打印出来。

26. If you want to see the art in the computer museum, what do you do?

【解析】细节考查题。文中提到，你根本就不需要亲自去博物馆，你只需要一台计算机和电话连在一起就可以了。

27. The director says why they started the museum?

【解析】细节推理题。文中提到，Robert Shostak 说过他建立这个图书馆的原因是计算机艺术家们没有地方展示他们的作品。

28. What can a laser printer do?

【解析】细节考查题。文章最后提到，艺术家们若想把自己的作品打印出来就必须安装一个激光打印机。

Passage Two

Men and Women in the United States who want to become doctors usually attend four years of college or university; next they study for four years in a medical school. After that they work in hospitals as medical residents or doctors in training. Some people study and work for as many as 13 years before they begin their lives as doctors.

During their university years, people who want to become doctors study science intensively. They must study biology, chemistry and other sciences. If they do not, they may have to return to college for more education in science before trying to enter medical school.

There are 125 medical schools in the United States. It is difficult to gain entrance to them. Those who do the best in their studies have a greater chance of entering medical school. Each student also must pass a national examination to enter a medical school. Those who get top score have the best chance of being accepted. Most people who want to study medicine seek to enter a number of medical schools. This increases their chances of being accepted by one. In 1998, almost 47,000 people competed for about 17,000 openings in medical schools.

Questions 29 to 31 are based on the passage you have just heard.

【内容概要】本文讲述的是在美国从医之路的艰难。在美国，任何人想从事医生的职业必须经过"4 年普通大学+4 年医学院+1～5 年的实习期"的洗礼。入学的竞争相当激烈，而在学院里，他们必须深入地学习生物学、化学、医药学等各门科学。

29. How many years does a student have to study before beginning his life as a doctor?

【解析】细节考查题。文章开始就提到，在美国，任何想从事医生职业的人必须经过"4 年普通大学+4 年医学院+1～5 年的实习期"，从 9 年到 13 年不等。

30. Which subject do the students who want to become doctors not study?

【解析】细节推理题。文中提到，想从医的学生要学习各门科学学科，但没提到哲学。

31. Which of the following is not true about the entrance to the medical schools?

【解析】综合推理题。医学院的入学竞争很激烈，优秀的学生有各种优势，但不能说对所有人而言机会都很多。

Passage Three

The Coast Guard does what its name says: it guards the coasts of the United States. During a war, the Coast Guard

118

becomes part of the United States Navy, and helps to protect against enemy attacks. In times of peace, however, the Coast Guard is part of the United States Department of Transportation. It has responsibility for many different duties. The Coast Guard can be found at many large lakes in America, as well as in coastal waters. It enforces laws controlling navigation, shipping, immigration, and fishing. It enforces other laws that affect the thousands of privately-owned boats in the United States.

Coast Guard planes, boats and helicopters search for missing boats and rescue people in dangerous situations. Last year, Coast Guardsmen saved the lives of almost 7,000 people. The Coast Guard does scientific research on the ocean. It also uses ice-breaking boats to clear ice from rivers or lakes, so boats can travel safely. One of the Coast Guard's most important duties now is helping to keep illegal drugs out of the United States. Coast Guard boats, armed with guns, use radios and radar to find boats that may be carrying drugs. They stop the boats suspected of carrying drugs and search them. They seize the drugs and arrest the people on the boats. Last year, Coast Guardsman seized more than 800,000 kilograms of marijuana and cocaine. And they arrested more than 700 persons trying to bring illegal drugs into the United States.

This kind of action is exciting. Most of the time, however, Coast Guardsman say they see nothing more exciting than the ocean.

Questions 32 to 35 are based on the passage you have just heard.

【内容概要】美国海岸警卫队是美军后备役部队，担负着战时与和平时的双重使命。和平时期，它属于运输部，而战时它则是海军的一部分。

32. What does the name Coast Guard mean?

【解析】细节考查题。文章第一句就提到美国海岸警卫队的任务就是保卫美国的海岸。其他几个选项都太片面。

33. Which of the following is not the duty of the U.S. Coast Guard?

【解析】细节综合题。文中提到在和平时期美国海岸警卫队负责有关航海、运输、移民、私人船舶等法律的实施以及营救遇难船只人员，但没提到训练人游泳。

34. How many lives have Coast Guardsman saved last year?

【解析】细节考查题。本题主要是测验学生对数目的掌握，尤其是一些在语音上比较难辨的数字。

35. What does the speaker indicate about the life of the Coast Guardsmen for most of the time?

【解析】综合推理题。尤其是让学生学会理解言外之意。文章最后说尽管前面提到的美国海岸警卫队的生活看似紧张兴奋，但大多数时间他们面对的只是一片大海，应该是枯燥乏味的。

Section C

The (36) birthplace was open at 10 a.m. It was 9:30 and already the pilgrims had formed a(37) queue I asked a lady from Ohio why she had come. "For Shakespeare," she said. "Isn't that why you came?"

"Not (38) entirely " I said. "I was born here. I'm visiting my family."

"You were born here?" she said, as if only Shakespeare had the right.

It was my first time in many years. Long ago I had (39) emigrated to America. Now I was visiting places in which I had taken little (40) interest before: the birthplace for example. I had passed it perhaps a hundred times without a thought of going in. Now it would cost me just under two pounds, about $3. An even stranger experience was buying a ticket to the school two(41) members of my family had attended. Shakespeare had gone there, though 350 years before. It was a good school, but I was (42) fortunate in being sent to a better one. "Better than Shakespeare's?" asked an

American to whom I had confided. "I don't see how that could be(43) possible", he had muttered before turning away.(44) I had taken Shakespeare for granted. However, in my current tourist status, that would have to be changed. (45) Respect was called for. I must learn to refer to him as "the bard", and not as "Will" in the familiar way, and never a "Willie the Shake", which is the inelegant but customary nickname of some of the younger generation. This was no problem. Shakespeare worship had begun before my day. Every building with Shakespeare connections was preserved. (46) An inclusive ticket enabled holders to see them all.

When several Americans whom I had run into asked me to show them around, I readily agreed.

Part IV   Reading Comprehension (Reading in Depth)
Section A

【短文大意】对于公众演讲者来说,如何表达自己和与别人交际的能力是尤其重要的,对于那些想在充满竞争的社会里立足的人这点也是十分重要。我们怎样才能获得这种能力,尤其是对于那些生来就胆小和害羞的人?

首先,认为公开演讲很重要的观点毫无疑问是正确的。无论他们是否和你一样将成为一名记者,或是否将成为一名商人,或是否将接受教育然后成为一名教师,人们都需要很好地表达自己。而且如果有人打算学习法学,他们必须是能有效地与别人交流,并且许多交流都要采取公开演讲的形式。有的交流是采取书面的形式,比如新闻工作。我们有出版印刷新闻和播报新闻,但是有效交流的原则基本上是相同的。当然,那些都是交流的形式。所以,如果你能学会一个优秀公众演讲者的所有技能,你不但能把这些技能运用到演讲中,而且能运用到其他形式的交流上。在写作和对话中你就能运用这些技能。几天前我与新华社的一个记者交谈,他说:"正如你所知,你现在所说的恰好和我们在文章和新闻中做的是相同的事情。你要有一个线索,引起读者的注意;要有一个主要的观点,还要继续这个主要的观点;要有第二个观点,要继续这个观点,最后要有归纳总结的结论"。所以公开演讲的技能是可以被应用到其他交流领域或者和其他交流领域关联起来。

47.【答案及解析】F) 空格处的词应该是动词,因为前面有不定式符号 to,所以只能选动词原形的动词。通过上下文来看符合题意的只有 F,to establish oneself in a society 是固定的短语,含义是"在社会立足"。

48.【答案及解析】E) 空格处很明显是形容词,修饰 competition。能够修饰 competition 的形容词只有 fierce(激烈的),因此 E)为正确选项。

49.【答案及解析】N) 在代词 your 之后只能是名词或相当于名词的动名词,符合这一条件的选项有几个,但从上下文语意来看,只有 point(观点,看法)才符合,因此 N)为正确答案。

50.【答案及解析】I) 在 be able to 后只能是动词。根据上下文,express oneself(表达自己)才符合语意,因此选择 I)。

51.【答案及解析】H) 在系动词 are 的后面,一般应该接形容词、现在分词或过去分词。选项后是 education,只有选 going in for,该短语表示"参加",因此答案为 H)

52.【答案及解析】L) 从上下文来看,从事法律的人应当能够 express themselves effectively all the time,但 express 已在前面用过了,不能再用。communicate 可以用作及物动词或用作不及物动词,故选 L)。

53.【答案及解析】J) 空格处需要填一个名词。从上下文来看,前文讲的是口头表达,如老师与律师,接下来讲书面表达,如新闻记者的文章,这两者在实际"有效"上其"原则"应该是一致的,因此选 J)。

54.【答案及解析】K) 此处是固定表达。新华社应该用 Agency,而不是 Association,故答案是 K)。

55.【答案及解析】D) 原文引述的是一位记者讲有效交流的经验。我们说话或写文章除了标题醒目之外,还要吸引听众或读者的注意力,即 get 或 draw the readers' attention,故选 D)。

56.【答案及解析】B) 这里是一个固定短语。文章的结尾当然要对全文进行归纳总结，这就是 sum up，因此 B) 为正确选项。

Section B

Passage one

In contemporary Asia most countries have granted equal status to women legally. However, in many countries the social mixing of men and women is still viewed with distrust and a woman is expected to remain in the background. If a woman has a profession, it is almost as if she is an abnormal kind of woman.

同义转述

Fortunately, social attitudes do not last forever. Although the change is slow, there is some change in Asian opinion about women having jobs. Medicine, nursing and teaching have the longest history. Women doctors were necessary in some countries where male doctors were not permitted, for social or religious reasons, to see women patients. Another recognized profession for women is to work in offices and some have reached very high positions. This tends to suggest that in some areas men's refusal to accept the professional ability—and equality—of women is gradually being eroded.

归纳综合

But being allowed to have a job is not enough. True liberation can't exist until there is wider social equality. This is not a one-way process. Women need to be educated to understand the meaning of their right, particularly in countries where the status of women has been low in the past. Yet in just these countries a widely held belief is that if women are educated they will become less "womanly".

In many ways the fight for the real liberation of women throughout Asia has only just begun. Although women now have political and legal rights, practicing these rights in everyday life is a far more difficult matter. The frequently heard remark "You've got your rights now, what more do you want?", sums up this feeling. Asian women now want to put these rights into practice and it is here that they meet with opposition. "Women", one hears men say, "are getting too forward these days."

57. The so-called "normal" women should_____.
    A) be independent and educated
    B) stay at home all the day
    C) stay in the background socially
    D) have a stable and good job

58. Medicine, nursing and teaching are jobs_____.
    A) recognized for women
    B) for men and women
    C) beyond women's ability
    D) for professional women

59. The word "eroded" in Para.2 probably means_____.
    A) strengthened
    B) recognized
    C) reduced
    D) changed

词义推理

60. What's the author's attitude towards women's liberation?
    A) Indifferent
    B) Supportive
    C) Disappointed
    D) Humorous （归纳推理）

61. What's the main idea of this article?
    A) Asian women have gained their rights.
    B) There are many good jobs for women.
    C) Asian women want to put their rights into practice.
    D) The liberation of Asian women. （主旨大意）

121

【短文大意】在当代的亚洲，大部分国家都能够在法律上给予妇女平等的社会地位。然而，在很多国家，女人和男人共同立足于社会依然得不到信任，并且妇女被期望留在幕后。如果一个女人拥有自己的职业，这似乎意味着她是一个不正常的人。

幸运的是，这种社会态度不会永远持续下去。尽管变化比较缓慢，但是亚洲人对于职业妇女的看法有了一定程度的变化。女性在医学、护理、教育领域就职有着最悠久的历史。在某些国家，由于社会和宗教的原因，男医生不允许检查女病人，女性医生是很必要的。另一个被认为属于妇女的职业是在办公室工作，并且有一些妇女已达到了很高的职位。这种趋向表明在一些地区，男人拒绝接受女人的职业能力和职业平等这种观点正逐渐消除。

但是，仅仅允许妇女拥有工作还是不够的。只有当妇女享有更广泛的社会平等时才会存在真正的解放。这不是一个单向的过程。妇女需要接受教育来理解她们权利的涵义，尤其是在那些过去妇女地位很低的国家。然而，就是在这些国家，人们普遍认为，如果一个女人受教育，她就会变得缺少女人味。

在许多方面，遍及整个亚洲，为了争取真正的妇女解放的斗争才刚刚开始。尽管妇女现在拥有了政治和法律上的权利，但将这些权利在日常生活中付诸实践还是一个难题。经常听到"你已经得到你的权利了，你还想要什么？"这样的话，这足以说明这种现状。现在，亚洲的妇女想把这些权利付诸实践，正是在这点上她们遭到了反对。曾有人听男人说，"女人，现在获得的太多了。"

57.【题目译文】所谓正常的妇女应该_____。

【答案及解析】C) 这道题是就第一段进行提问的。答题关键在于"... is still ... to remain in the background"。在许多国家，妇女被期望留在幕后。如果一个女人拥有一个职业，这似乎意味着她是一个不正常的人。由此可知最符合题意的是C)。

58.【题目译文】医学、护理和教育是一些_____的工作。

【答案及解析】A) 这道题是就第二段进行提问的。"Medicine, nursing ... the longest history."这句话表明医学、护理和教育是大家均认可的有着悠久历史的职业，且下文中"women doctors were necessary ..."和"Another recognized profession for women ..."指出"由于社会和宗教的原因，女医生在某些国家是很必要的。另一个被认为属于妇女的工作是……"，由此可以确定A）项是正确答案。

59.【题目译文】第二段中"eroded"的含义是_____。

【答案及解析】C) 这道题是针对第二段最后一句进行提问的。由上文可推断，男性拒绝女性参加工作的态度正在渐渐变得缓和，因而选C)。

60.【题目译文】作者对妇女解放的态度是什么？

【答案及解析】B) 从整篇文章来看，作者一直都抱着支持理解的态度，而且十分关注女性解放的问题，因而选B)。

61.【题目译文】这篇文章的主要观点是什么？

【答案及解析】D) 主旨题。这道题是就整个语篇进行提问的。尽管亚洲妇女现在拥有了政治和法律上的权利，但将这些权利付诸实践还是一个难题。在许多方面，遍及整个亚洲，为了争取真正的更广泛的社会平等的妇女解放的斗争才刚刚开始。通篇都在谈论亚洲妇女的解放。因此D)项最为合适。

Passage Two

With a tremendous roar from its rocket engine, the satellite is sent up into the sky. Minutes later, at an altitude of 300 miles, this tiny electronic moon begins to orbit about the earth. Its radio begins to transmit an astonishing amount of information about the satellite's orbital path, the amount of radiation it detects, and the presence of meteorites. Information of all kind races back to the earth. No human being could possibly copy down all these facts, much less remember and organize them. But an electronic computer can.

The marvel of the machine age, the electronic computer has been in use only since 1946. It can do simple computations—add, subtract, multiply and divide—with lighting speed and perfect accuracy. Some computers can work 500,000 times faster than any person can.

Once it is given a "program" — that is, a carefully worked-out set of instructions devised by a technician trained in computer language—a computer can gather a wide range of information for many purposes. For the scientist it can get information from outer space or from the depth of the ocean. In business and industry the computer prepares factory inventories, keeps track of sales trends and production needs, mails dividend checks, and makes out company payrolls. It can keep bank accounts up to date and make out electric bills. If you are planning a trip by plane, the computer will find out what to take and what space is available.

Not only can the computer gather facts, it can also store them as fast as they are gathered and can pour them out whenever they are needed. The computer is really a high-powered "memory" machine that "had all the answers" —or almost all. Besides gathering and storing information, the computer can also solve complicated problems that once took months for people to do.

At times computers seem almost human. They can "read" hand-printed letters, play chess, compose music, write plays and even design other computers. Is it any wonder that they are sometimes called "thinking" machines?

Even though they are taking over some of the tasks that were once accomplished by our own brains, computers are not replacing us at least not yet. Our brain has more than 10 million cells. A computer has only a few hundred thousand parts. For some time to come, then, we can safely say that our brains are at least 10,000 times more complex than a computer. How we use them is for us, not the computer, to decide.

62. In the first paragraph, the author thinks an electronic computer can_____.
   A) copy down all the facts
   B) remember all the facts
   C) organize the facts and everything
   同义转述 D) copy down, remember and organize all the facts

63. "Program" means_____.
   A) a plan of what is to be done
   B) a complete show on a TV station at a fixed time table
   C) a scheduled performance
   同义转述 D) series of coded instructions to control the operations of a computer

64. The computer is a high powered "memory" machine, which_____.
   A) has all the ready answers—or almost all to any questions
   B) can remember everything
   C) has all the answers—or almost to all the information that has been stored
   D) can store everything and work for you

65. "Thinking" machines suggest that_____.
   A) they can "read" hand printed letters etc.
   同义转述 B) they can't think, but can do something under human control
   C) they even design other computers
   D) they really can think and do many other jobs

66. Why can't computers do whatever they want to do?
   A) Because some computers can't work 500,000 times faster than any person can.
   B) Because they normally have a few hundred thousand parts.
   C) Because human brains are at least 10,000 times more complex than any computers.
   同义转述 D) Because how a computer works is decided by human.

【短文大意】随着火箭引擎的巨大呼啸，卫星被发射进入太空。几分钟后，在海拔达 300 英里的地方，这个小型的电子月亮开始环绕地球运转。它通过无线电开始传输出关于卫星运转轨道、所探测到的辐射的数量以及陨石分布等令人惊奇的信息。各种各样的信息被发射到地球上。没有一个人能把这些数据复制下来，更不用说记住并把它们组织起来，但是计算机做到了。

机器时代的一个奇迹就是 1946 年电子计算机的首次应用。那时计算机就能以闪电般的计算速度和极高的准确度进行简单的计算——加、减、乘、除。某些计算机能比任何人的工作速度快 50 万倍。

一旦计算机被安装上一套程序——也就是一套由技术人员使用计算机语言编制出的工作指令——计算机就能存储大范围的有着不同用途的信息。它能为科学家获得来自太空的或来自大洋底部的信息。在工商业领域，它能为工厂发明做准备，能及时跟踪销售趋势以及产品需求、邮办红利现金和核算公司工资。它还能更新银行账户，也能结算电费。如果你打算乘飞机旅行，计算机能帮你查明乘哪个班机和买到什么位置的票。

计算机不仅能收集信息，也能尽快地存储信息，并且能在任何时候调出需要的信息。它确实是一个能够存储所有问题答案的、有着超强实力的记忆机器。除了收集存储信息，它还能解决曾经困扰人们数月的复杂问题。

有时候，计算机看起来几乎就是人类。它们会读打印的信件，会玩象棋，会谱写音乐，会写剧本甚至会设计其他计算机。难怪有时候它们会被叫做"善于思考"的机器。

即使计算机接管了一些曾由我们的大脑成功完成的任务，它们也不会取代我们，至少现在还没有取代我们。我们的大脑有 1000 多万个细胞。而计算机仅仅有几百万个零件。所以在将来很长的一段时间内，我们可以非常确信地说，我们的大脑比计算机至少要复杂 1 万倍。如何使用计算机是由我们而不是由计算机来决定的。

62. 【题目译文】在第一段中，作者认为电子计算机能够_____。

【答案及解析】D) 此题为段落大意理解题。计算机不仅能复制，又能存储，还能排序。由文章第一段最后两句话可以确定 D)为正确答案。

63. 【题目译文】program 的含义是_____。

【答案及解析】D) 从原文第三段第一句话 "Once it is given a "program" —that is, a carefully worked-out set of instructions devised by a technician trained in computer language…"，可知程序就是一套由技术人员使用计算机语言编制出来的，用来控制计算机操作的工作指令。由此可以确定 D)为正确答案。

64. 【题目译文】计算机是一个超强的记忆机器_____。

【答案及解析】C) 计算机提供有关它所存储的信息的答案。另外，由第四段第二句话 "The computer is really a high-powered "memory" machine that "had all the answers" —or almost all." 即：它确实是一个能够存储所有问题答案的有着超强实力的记忆机器。因此可以确定正确答案为 C)。

65. 【题目译文】"善于思考"的机器表明_____。

【答案及解析】B) "善于思考"的机器并不是指计算机能够思考，而是计算机在人类的控制下完成指定的任务。由原文倒数第二段可以推断出 B)符合题意。

66. 【题目译文】为什么计算机不能做任何它们想做的事？

【答案及解析】D) 见原文最后一段最后一句话 "How we use them is for us, not the computer, to decide."，即：如何使用计算机是由我们而不是由计算机来决定的。同时根据常识计算机是按人类安装的工作指令，即计算机程序来工作的，因此，D)为正确答案。

Part V   Cloze

【译文】 一个夏日的夜晚，在下班回家的路上，我决定看一场电影。我知道电影院有空调，而我无法面对

124

我那酷热的公寓。

坐在电影院里，我得透过前面两个人的脑袋中间的空隙看电影。每当那位女士倾过身和男友说话或者男士倾过身来亲吻女友时，我就得改变视线的角度。为什么美国人要在公共场所表达这样的情感？

我本想那部电影会对我的英文有好处，结果却发现是一部意大利片子。过了大约半个小时，我决定不看电影了，集中精力吃爆玉米花。我至今也没搞明白为什么他们给我那么多爆玉米花，不过，吃起来倒还不错。过了一会儿，我听不到意大利人浪漫的对话，却只听到爆玉米在牙齿间的咀嚼声音。我的思绪也开始漫游起来。我记得，在韩国的时候，我常常在电视上看 kojak 节目，他韩语讲得很标准真令我惊讶，他看起来像是我的一个好朋友，直到有一天，我在纽约又一次遇到了他，他说着相当流利的英文，而不是那流利的韩语，甚至没有一丝韩语的口音，那时，我才发现自己被骗了。

六年前，我们全家移民到了美国，在那时，我们没有一个人会说英文。后来，我们开始学习一些单词，妈妈建议我们在家也应该说英语，大家一致同意，但是我们家变得非常安静，我们似乎相互回避，我们静静地坐在桌前，宁愿一言不发也不用一种难说的语言交流，母亲试图想用英语说点什么，结果错误百出，我们都禁不住大笑起来，然后决定不说英语了。从那以后，我们在家都说韩语。

67.【答案】A 根据题目意思，夏天没有空调，公寓应是很热的，因此他去剧院，一个有空调的地方。

68.【答案】B 这里指的是两个人脑袋之间的空隙。Crack 裂缝，缝隙, opening 口，孔 e.g. he put a gate across the opening in the fence. 他在围墙的开口处安了一个门。Break 破裂，缝隙 e.g. a break in the clouds 云多间的一线青天。Blank 空白的 e.g. write your name, address and telephone number in the blank spaces at the top of the page. 在这一页顶上的空白处写上你的姓名、地址和电话号码。根据题目意思，应该选择 opening。

69.【答案】B 根据上下文，当前面的女子不断斜过身与男子说话时，我当然要不断改变角度才能看到银幕。表示角度的词是 angle。

70.【答案】D "女的侧身过去与男的说话"与"男子侧身吻女子"两种情况在此交替出现，且应该是并列关系，所以，要选并列连词 or。

71.【答案】B 能与 display "显示，展示"搭配的是 affection "情感或感情"。本句意思为：美国人为什么要在公共场所表达这样的情感呢？

72.【答案】C 此句中 as 引导一个定语从句，其先行词是后面的 it was an Italian movie. as 可以翻译成为：正如…

73.【答案】A 从时间关系上来看，作者是看了一个小时后才决定放弃，所以答案是 after.

74.【答案】D 根据上下文中提到，concentrate on 是一个固定搭配，意为"专心于……"用在此处最恰当。

75.【答案】B It tasted pretty good 与上文的 I've never understood why they give you so much popcorn 之间是让步关系，因此选择 though。

76.【答案】A 根据下文 I just heard the sound of the popcorn crunching between my teeth（我只听到嚼玉米花的声音）可知，他没再听到意大利语的浪漫声音。

77.【答案】C sound 泛指声音，voice 特指发出的声音。

78.【答案】D 从下文中他想的内容可判断，他的思绪开始游荡（wander）. wander …对感到惊讶，惊奇，想知道；depart 离开，启程 e.g. when does the next train depart? 下一次列车什么时候开？；Imagine 想像。

79.【答案】A 因为他是在叙述过去的事情，又表示经常性的行为，应该用 used to do sth.。

80.【答案】B 从下文中的 betray 可以判断，他在发现这一情况之前一直把对方当作朋友，发现真实情况之后才发现自己被背叛了，Until 表示"在……之前"。

81.【答案】D 根据下句，讲英语时一点儿韩国人的口音都没有，应该是完美的，再照应前一句的 perfect Korean,

所以选 perfect.

82.【答案】B 根据句义，他感觉自己好像被背叛了。Feel 表示"感觉"。

83.【答案】C 本句意思为：我们一开始学习英语，妈妈就提出了一个建议，建议我们在家里都说英语。Once 表示"一……就……"。

84.【答案】A 从下文中的 we all seemed to avoid each other 与 we sat at the dinner table in silence 可知，答案是 quiet，意思是大家都保持沉没，屋里十分安静，吃饭时也都是默默地吃。

85.【答案】D prefer A to B 意思是 like A better than B，to 是介词。A，B 和 C 都是及物动词，只有 speak 是不及物动词，与 in 搭配，speak in a difficult language "用一种很难的语言讲话"。

86.【答案】D 这句话的意思是：妈妈试着说点英语，结果是错误百出，我们忍不住放声大笑。Come out "结果是……"；work out "可以解决，设计出，作出，计算出"；get out "出去，离开，逃脱，泄露"；make out "设法应付，理解，辨认出，填写"。

Part VI   Translation

87. The substance does not dissolve in water_____(不管是否加热)。

【题目译文】这种物质不能在水中溶解，不管是否加热。

【答案及解析】whether(it is) heated or not   注意条件句的使用。

88. There are several people like me who would _____(以某人为榜样) over anything else.

【题目译文】这有许多像我一样的人以某人为榜样胜过了一切。

【答案及解析】follow your lead over   胜过，超出。

89. You should _____(抽出时间和力量来帮助别人)when necessary。

【题目译文】在必要的时候，你应该抽出时间和力量来帮助别人。

【答案及解析】give of yourself   注意 when（it is）necessary 中省略部分的使用。

90. You are a popular girl, Linda, and many younger ones _____(尊敬)。

【题目译文】琳达，你是一个受欢迎的女孩，受许多年轻人的尊敬。

【答案及解析】look up to   注意 look up to 除有"向上看……"的意思外，还有"尊敬（人、行动）"的意思。

91. No one could _____(代替) her father。

【题目译文】没有人能够代替她的爸爸。

【答案及解析】 "take the place of"有"代替"的意思。

大学英语4级考试新题型模拟试卷八答案及详解

参考答案

Part I   Writing

Knowledge and Certificates

People usually connect knowledge with certificates. When you have got a certain amount of knowledge in a school or a university, you can get a certificate. A certificate is a ruler to measure whether a person is qualified after studying for a period of time.

It doesn't mean, however, a person who has rich knowledge should have a certificate. A person can get knowledge by self-studying or in practice. So a certificate can not be used to measure a person's ability for a certain job.

And we can not take it for granted that a person with a certificate is knowledgeable. Some people make false certificates. Some people buy certificates with money. This action cheats not only the public society, but also the buyers themselves. In order to spread actual knowledge, the government should make laws to prevent the spreading of cheating.

Part II   Reading Comprehension (Skimming and Scanning)

1. Y   2. Y   3. Y   4. NG   5. N   6. Y   7. NG

8. a variety of factors   9. the police will arrive in five minutes   10. establishing realistic expectations

Part III   Listening Comprehension

Section A

11. B   12. C   13. D   14. D   15. C   16. A   17. C   18. B   19. B   20. B
21. C   22. C   23. B   24. B   25. C

Section B

26. C   27. C   28. A   29. A   30. C   31. B   32. D   33. B   34. D   35. D

Section C

36. aim   37. genius   38. disaster   39. educational
40. aware   41. damage   42. succeed   43. supportive

44. He is crazy about music, and his parents help him a lot by taking him to concerts and arranging private piano and violin lessons for him

45. Because both his parents are successful musicians, they set too high standard for Winston

46. "When I was your age, I used to win every competition I entered."

Part IV   Reading Comprehension (Reading in Depth)

Section A

47. H: aggressively   48. O: penalties   49. M: ruining   50. D: exposure   51. G: awarded
52. N: valid   53. F: unqualified   54. C: bought   55. L: list   56. I: typing

Section B

57. C   58. A   59. B   60. B   61. A   62. B   63. D   64. C   65. A   66. B

Part V   Cloze

67. D   68. A   69. C   70. B   71. A   72. C   73. D
74. B   75. A   76. C   77. B   78. A   79. D   80. C
81. B   82. A   83. C   84. B   85. D   86. D

127

87．have a fit   88．came into being   89．you name it

90．left off   91．drawn someone's attention

答案解析及录音原文

Part I   Writing

Knowledge and Certificates

People usually connect knowledge with certificates. When you have got a certain amount of knowledge in a school or a university, you can get a certificate. A certificate is a ruler to measure whether a person is qualified after studying for a period of time.

It doesn't mean, however, a person who has knowledge should have a certificate. A person can get knowledge by self-studying or in practice. So a certificate can not be used to measure a person's ability for a certain job.

And we can not take it for granted that a person with a certificate is knowledgeable. Some people make false certificates. Some people buy certificates with money. This action cheats not only the public society, but also the buyers themselves. In order to spread actual knowledge, the government should make laws to prevent the spreading of cheating.

【评论】本文是一篇议论文，针对当今社会所关注的问题：知识和文凭之间的关系，进行了议论。本文论点明确，论据有理有据，逻辑性强，是一篇很好的文章。

【常用句型】   第一段：People usually connect… with…

第二段：So a certificate can not be used to measure a person's ability for a certain job.

第三段：And we can not take it for granted that a person with a certificate is knowledgeable.

Part II   Reading Comprehension (Skimming and Scanning)

【文章及答案解析】本文主要讲的是《快速的警察反应》，有点类似于我国的 110。通过阅读，可知文章包括快速反应的时间定义、应用方法、应用时机以及长处和弊端等内容。

Rapid Police Response

A

Police departments in the United States and Canada see it as central to their role that they responds to calls for helps as quickly as possible. This ability to react fast has been greatly improved with the aid of technology. The telephone and police radio, already long in use, assist greatly in the reduction of police response time. In more recent times there has been the introduction of the "911" emergency system, which allows the public easier and faster contact with police, and the use of police computer system, which assist police in planning patrols and assigning emergency request to the police officers nearest to the scene of the emergency.

【文章大意】

快速警察反应

A

美国和加拿大的警方认为对接到的求助进行快速反应是职责的关键。快速反应能力在科技的辅助下已得到了很大的提高。早已长期使用的电话和警讯电台，极大地帮助了警察缩短反应时间。近期推出的"911"紧急系统，可以让市民更加方便快速地联系警察，此外警局内电脑系统的使用有助于警方制定巡逻计划和分配应急任务给离紧急事发地最近的警察。

B

As an important part of police strategy, rapid police response is seen by police officers and the public alike as offering tremendous benefits. The more obvious ones are ability of police to apply first-aid life-saving techniques quickly and the greater likelihood of arresting people whomay have participated in a crime. It aids in identifying those who witnessed an emergency crime, as well as in collecting evidence. The overall reputation of a police department, too, is enhanced if rapid response is consistent, and this in itself promotes the prevention of crime. Needless to say, rapid response offers the public some degree of satisfaction in its police force.

C

While these may be the desired consequences of rapid police response, actual research has not shown it to be quite so beneficial. For example, it has been demonstrated that rapid response leads to a great likelihood of arrest only if responses are on the order of 1~2 minutes after a call is received by the police. When response times increase to 3～4 minutes—still quite a rapid response — the likelihood of an arrest is substantially reduced. Similarly, in identifying witnesses to emergencies or crimes, police are far more likely to be successful if they arrive at the scene no more than four minutes, on average, after receiving call for help. Yet both police officers and the public defined "rapid response" as responding up to 10～12 minutes after calling the police for help.

D

Should people police assume all the responsibility for ensuring a rapid response? Studies have shown that people tend to delay after an incident occurs before contacting the police. A crime victim may be injured and thus unable to call for help, for example, or no telephone may be available at the scene of the incident. Often,

B

快速反应作为警务战略的一个重要组成部分，被警方和广大市民看成为能够带来巨大益处的方法。其中最明显的是快速使用急救用的救生设备的能力和最大可能逮捕参与犯罪人员的能力。

这有助于对紧急情况或犯罪现场目击者的确认及证据的收集。如果快速反应一致，警局的整体声誉也会得到提高，而且这本身也是对犯罪预防的推动。不必说，快速反应带给市民的是对警察机关的某种满意程度。

细节推理

1. Police believe there is a better chance of finding witnesses to a crime if response is rapid.(Y)

细节推理

2. A response delay of 1~2 minutes may have substantial influence on whether or not a suspect criminal is caught. （Y）

3. The public and the police generally agree on the amount of time normally taken for a rapid response. (Y)

C

或许这是对警察快速反应结果的渴望，但实际上研究表明它并不完全有成效。例如，只有反应时间在警察接到电话后的1～2分钟，快速反应才会使抓住罪犯的可能性更大。当反应时间增加到3～4分钟，仍然是非常快的反应，抓住罪犯的可能性就实质性地降低了。同样，假使在接到求救电话后平均不超过4分钟抵达现场，警察则更有可能成功地对紧急情况或犯罪案件的目击者进行确认。然而无论是警方还是市民，他们对"快速反应"的定义是警察在接到求救电话后10～12分钟内作出的反应。

D

警察应该承担确保快速反应的全部责任吗？调查显示人们倾向在事故发生后拖延与警察联系，例如，

细节确认

4. Physical barriers are the greatest cause of delay in contacting police. (NG)

however, there is no such physical barrier to calling the police. Indeed, it is very common for crime victims to call their parents, their minister, or even their insurance company first. When the police are finally called in such case, the effectiveness of even the most rapid of responses is greatly diminished.

受害者可能因伤无法打电话求救，或事故现场的电话无法使用，然而，通常来讲不是由于这种身体原因阻碍给警察打电话。的确，犯罪受害人给他们的父母、经理，甚至是保险公司打电话是最为平常的。当警察最后接到这样的电话时，即使是做出最快速的反应其效果也大大降低。

句法翻译

> 5.Rapid response is considered desirable in handling cases of burglary. （**N**）

E

The effectiveness of rapid response also needs to be seen in light of the nature of the crime. For example, when someone rings the police after discovering their television set has been stolen from their home, there is little point, in terms of identifying those responsible for the crime, in ensuring a very rapid response. It is common in such burglary or theft cases that the victim discovers the crime hours, days, even weeks after it has occurred. When the victim is directly involved in the crime, however, as in the case of a robbery, rapid response, provided the victim was quickly able to contact the police, is more likely to be advantageous. Based on statistics comparing crimes that are discovered and those in which the victim is directly involved, Spelman and Brown (1981) suggest that three in four calls to police need not be met rapid response.

E

快速反应的效果也需要根据案件的性质来衡量。如某人回家发现电视机被盗后打电话给警察这样的小事，应确认物主对此事应负的责任，以确保其反应的迅速。有些入室行窃、偷盗的案件，受害人是在案发后的几小时，几天甚至是几个星期才发现，这种情况屡见不鲜。然而当受害人被直接卷入到犯罪中，如遇上抢劫，倘若受害人能够快速地与警察联系，那么警察的快速反应则可能更为有效。以事后发现的案件和受害人直接卷入的案件相比较的数据为基础，史伯曼和布朗在1981年提出，在四个求救电话中有三个没有必要采取快速反应。

细节确认

> 6. Research shows that some 75% of crimes are discovered by victims after they have been committed. (**Y**)

F

It becomes clear that the importance of response time in collecting evidence or catching criminals after a crime must be weighed against a variety of factors. Yet because police department officials assume the public strongly demands rapid response, they believe that every call to the police should be met with. Studies have shown, however, that while the public want quick response, more important is the information given by the police to the person asking for help. If a caller is told the police will arrive in five minutes but in fact it takes ten minutes or more, waiting the extra time can be extremely frustration. But if a caller is told he or she will have to wait 10

> 8. The importance of response time in collecting evidence must be weighed against_____.(**a variety of factors**)

F

很明显在权衡犯罪的各种因素后，收集证据和抓捕罪犯的过程中反应的时间是很重要的。然而由于警察部门的官员们承担着对市民强烈需求作出快速反应的职责，他们认为打给警察的每一个电话都应该处理。

细节推理

> 7. Police departments are usually successful in providing a rapid response regardless of the circumstances of the crime or emergency. (**NG**)

minutes and the police indeed arrive within that time, the caller is normally satisfied. Thus, rather than emphasizing rapid response, the focus of energies should be on establishing realistic expectations in the caller and making every attempt to meet them.

但研究表明,市民们越想让警察做出快速的反应,警方对求助者提供的信息就越重要。如果打电话者被告知警察将在5分钟内到达而实际上花费10分钟或更多,那么对多于时间的等待将是让人十分沮丧的。但当打电话者被告知他/她需等待 10 分钟,而警察确实在该时间内到达,打电话的人通常会很满意。因此,与其强调快速反应,不如将精力集中在对打电话者实际期盼的建立上,尽一切努力来满足他们。

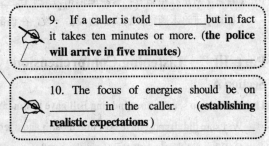

9. If a caller is told _____ but in fact it takes ten minutes or more. **(the police will arrive in five minutes)**

10. The focus of energies should be on _____ in the caller. **(establishing realistic expectations)**

1.【题意】警察认为如果能反应快速,就能加大寻找犯罪案件的目击者的机会。

【解析】本题的关键词为 witness,而不是 response 和 rapid,后者虽然很重要,但在全文很普及,连标题中也含有,因此不具有特征性。若想对本题定位,可见文章第二段的第三句话,但仅凭第三句是不行的,其首字母有 it,我们从这句话看起,才能很好地理解 it 在文中指代的是什么,从而推论出题目的命题。故选 YES。

2.【题意】1～2 分钟的反应延迟会对嫌疑犯是否被抓住产生实质性影响。

【解析】本题的关键词是1～2 minutes 可见原文 C 段中的第2、3句,其中说道"反应需要的过程为 1～2 minutes"和"如果反应时间增加到 3～4 mintues",也就是说在原来的基础上增加了 1～2 minutes,由此可见本题的说法是正确的,故此题选 YES。

3.【题意】大众和警察都普遍认为应该把一部分的时间用于快速反应。

【解析】从题干中的 "The public and the police" 可定位在 C 段的最后几句话,很明显本题是该句的总结陈述,而 the amount of time 正好对应 10～12 minutes,故此题选 YES。

4.【题意】身体的不适是造成与警方联系迟缓的最主要的原因。

【解析】本题的关键词是 barriers,很多人会误将 greatest 作为关键词。在第四段看到这样一句话 "often, however, there is no such physical barrier to calling the police"。很多人会选 NO,但需要注意的是,often no 不等于 never,既然存在,虽然不经常有,只是数量的问题,而 greatest(最大)则是程度上的问题,两者不具可比性,是典型的偷换概念的问题。故此题选 NOT GIVEN。

5.【题意】在入室抢劫方面,快速反应被认为是值得的。

【解析】本题考查的重点是在理解通过 burglary 定位出来的复杂句子上,burglary 在文中 E 段的第三句中,然而 burglary 前有个 such,表明上下文提及过在第二句的理解上要注意,第二句的意思与题意正好相反,故选择 NO。

6.【题意】研究表明约有 75% 的案件是受害者事后发现的。

【解析】本题的关键是要理解 E 段后部分的含义。Spelman 和 Brown 把事后发现的案件和受害者直接卷入的案件相比(当场发现),建立在此统计基础上,他们提出,在四个求救电话中,有三个没有必要采取快速反应。也就是说,三个电话是事后发现的。故本题选 YES。

7.【题意】除了犯罪和紧急情况外,警察局在提供快速反应方面很成功。

【解析】通过对"Police departments"的寻读,我们可以对应到 F 段第二句,第二句讲的是警察局的假设及愿望，而题目中陈述的是事实，故本题选 NOT GIVEN。

8.【题意】在收集证据中，快速反应的重要性可以由_____来衡量。

【解析】根据题干中的 must be weighted 及结合问题与文章顺序一致的关系，可将本题定位到最后一段，故此题的答案为 a variety of factors。

9.【题意】如果打电话者被告知_____，但事实上却要用 10 多分钟。

【解析】根据题干，考虑本题的顺序，很容易将其定位到文章最后一段的中间部分，本题是其中抽出来的一小句话，故可知此题的答案为 the police will arrive in 5 minutes。

10.【题意】对于打电话者，警察应把精力集中在_____。

【解析】可将本题定位到文章的最后一段，在该部分中可找到答案，答案为 establishing realistic expectations。

Part Ⅲ   Listening Comprehension
Section A

11. M: Did you see the diamond ring Bill gave to Linda?

  W: I sure did. It must have cost him an arm and a leg.

  Q: What did the woman say about the ring?

【答案及解析】推理判断题。男士说：“看见 Bill 给 Linda 的钻戒了吗？”女士说：“是的，一定是花了他一条胳膊和一条腿。”可知正确答案是 B) It was very expensive.很贵。

12. W: I heard that yesterday's basketball match had been wonderful.

  M: Wonderful? Which fool told you that?

  Q: What does the man think about the basketball match?

【答案及解析】推理判断题。女士说：“我听说昨天的篮球赛非常精彩。”男士说：“精彩？哪个傻瓜告诉你的？”可知正确答案是 C) The man doesn't think the match is wonderful at all.

13. W: Could you give me something for the pain? I didn't get to sleep until four o'clock this morning.

  M: This medicine is the strongest I can give you.

  Q: Where did this conversation most likely take place?

【答案及解析】推理判断题。 女士说：“给我点止疼药可以吗？我今天上午四点才睡。”男士说：“这个是我能给你的药效最强的药了。”可知正确答案是 D) In a drug store.在药店。

14. W: Why don't you look happy, Paul, when I tell you that good news?

  M: I'm glad to know it, but I am having trouble with my bad tooth.

  Q: Why doesn't Paul look happy?

【答案及解析】推理判断题。女士说：“保罗，当我告诉你那个好消息的时候，你为什么看起来不开心呢？”男士说：“我很高兴知道，但是我牙疼。”可知正确答案是 D) He is ill. 他生病了。

15. M: What do you think we should do about the mistake?

  W: Why not ask Tommy?

  Q: What can we learn from the conversation?

【答案及解析】推理判断题。男士说：“关于这个错误你认为我们应该怎么做？”女士说：“为什么不去问汤米？”可知正确答案是 C) It is Tommy who made the mistake. 是汤米犯的错。

16. W: Well, if you are seriously considering buying a car, I'm trying to get rid of mine .All it needs is some new paint.

  M: Thanks. But most used cars end up being more trouble than their worth.

132

Q: What will the man probably do?

【答案及解析】推理判断题。女士说："如果你很认真地考虑买车，我会试图把我的卖掉，只需要喷漆。"男士说："谢谢。可绝大多数用过的车所带来的麻烦要超过它们本身的价值。"可知正确答案是 A) Buy a new car. 买新车。

17. M: I've been working on this report all day. And I've still got 10 pages to write. At this rate, I'll never get it done by tomorrow.

   W: Oh, that's right. You weren't in class today, so you probably haven't heard that the deadline has been extended a week.

   Q: What do we know about the man?

【答案及解析】推理判断题。男士说："我一整天都忙这个报告，还有 10 页没写，以这种速度，明天也写不完。"女士说："你今天没上课，所以你可能不知道最后期限延期一周。可知正确答案是 C) He will have time to finish his report. 他还有时间完成他的报告。

18. M: Betty, why isn't Ella teaching here this term?

   W: She can't. She was fired.

   Q: What reason was given for Ella's not teaching?

【答案及解析】推理判断题。男士说："贝蒂，这学期艾拉为什么不在这教学了？"女士说："她不能教了，她被解雇了。"可知正确答案是 B) She was dismissed from her job. 她被解雇了。

Now you will hear two long conversations.

Conversation One

W: Bob, can we really afford a holiday? We're paying for this house and the furniture is on HP and...

M: Now listen, Peggy. You work hard and I work hard. We're not talking about whether we can have a holiday. We're talking about where and when.

W: Shall we go to Sweden?

M: Sweden's colder than Sheffield. I'd rather not go to Sweden.

W: What about Florida? Florida's warmer than Sheffield.

M: Yes, but it's a long way. How long does it take to get from here to Florida?

W: All right. Let's go to Hawaii.

M: You must be joking. How much would it cost for the two of us?

W: But the brochure says the problem of money will disappear. Bob, where do you really want to go?

M: I'm thinking of Wales or Scotland .Do you know why?

W: Yes. They're right on our doorstep and so close to home.

Questions 19 to 21 are based on the conversation you have just heard.

19. What are the two speakers discussing?

   【解析】推理判断题。从对话的内容上看，Bob 和 Peggy 讨论的问题是度假的地点而不是动身的时间。

20. Why doesn't Bob want to go to Florida?

   【解析】事实细节题。当 Peggy 提出去 Florida 时，Bob 说：but it's a long way. 但是路途远。

21. Where does Bob want to go for the holiday?

   【解析】事实细节题。在排除几个地方后，Bob 说：I'm thinking of Wales or Scotland（我想去威尔士或苏格兰）。

Conversation Two

M: Math Department, Doctor Webster speaking.

W: Hello, Prof. Webster, this is Janet Hill calling, I'm living two doors down from your teaching assistant, Don Williams. Don asked me to call you because he has lost his voice and can't talk to you himself.

M: Lost his voice. Oh, what a shame! Is there anything I can do for him?

W: Well, he has a class this afternoon from two-thirty to four and he won't be able to teach it, but he doesn't want to cancel it either.

M: Want me to try to find somebody else to teach the class?

W: No, not exactly. What he wants to do is to get someone to go in for him, just to pass back the mid-term exams.—He's already marked them and they are on the desk in his office. The whole thing wouldn't take more than ten minutes.

M: His classes are two-thirty, Oh, Well, I'm afraid at that time I was going to be on campus anyway; so I could do it for him. What room is his class in?

W: Cader Hall, room two-fourteen, Will you need his office key to get the exams? He's given it to me and I could bring it to you.

M: Actually, that won't be necessary. We have a master key in the math department. So I can get in to his office.

W: Thank you very much, Prof. Webster. Don doesn't have another class to teach until Thursday, and hopefully, he will be able to talk by then. He'll call you as soon as he can. Oh, yes, I almost forgot. Could you put the next assignment on the board, too? It's all the problems on page forty-five, and they are due at the next class.

Questions 22 to 25 are based on the conversation you have just heard.

22. What is Don's problem?

【解析】事实细节题。对话中女士讲出了 Don 的麻烦: because he has lost his voice. 因为他失声了。

23. What favor does Don want someone to do for him?

【解析】事实细节题。Don 希望有人帮他把期中考试的试卷发给学生: just to pass back the mid-term exams。

24. What does Janet offer to do?

【解析】事实细节题。Don 的钥匙在 Janet 手里,由 "He's given it to me and I could bring it to you." 可知女士可以把钥匙送过来。

25. What does Janet almost forget to ask Professor Webster?

【解析】事实细节题。女士的最后一句话包含了本题的答案: I almost forgot. Could you put the next assignment on the board, too? (我差点忘记了,还要在黑板上留下次的作业。)

Section B

Passage One

When young people get their real jobs, they may face a lot of new, confusing situations. They may find that everything is different from the way things were at school. It is possible that they will fell uncomfortable in both professional and social situations. Eventually, they realize that university classes can't be the only preparation for all of the different situations that appear in the working world.

Perhaps the best way to learn how to behave in the working world is to identify a worker you admire and observe his behavior. In doing so, you'll be able to see what it is that you admire in this person. For example, you will observe how he acts in trouble. Perhaps even more important, you will be able to see what his approach to everyday situations is.

While you are observing your colleague, you should be asking yourself whether his behavior is like yours and how you can learn from his response to a different situation. By watching and learning from a model, you will probably begin to identify and get good working habits.

Questions 26 to 28 are based on the passage you have just heard.

【内容概要】本篇短文讲的主要是关于刚毕业的年轻人如何适应职场的问题。最好的办法是效仿一个你崇拜的人，看他的行为，或是观察你的同事，看你有没有和同事身上类似的毛病。

26. Why don't young people behave well in the working world after their graduation?

【解析】归纳题。从开始的前两句话：they may face a lot of new, confusing situations. They may find that everything is different from the way things were at school. （他们会面对一些新的形势，他们会发现任何事和他们在上大学的时候都不一样）。可知答案是 C) The society is too complicated.社会太复杂了。

27. What is the best way to learn to behave well in the working world?

【解析】细节题… is to identify a worker you admire and observe his behavior.（确定一个你尊敬的员工然后观察他的行为。）可知答案是 C) To find a person you respect and watch carefully how he acts.

28. Which is the best title for the passage you have just heard?

【解析】综合分析题。A) Learn From a Model 向一个榜样学习。

Passage Two

Bill Gates was born and brought up in Seattle. At the age of 14, he founded a computer programming company with three friends, and they had earned $20,000 by selling their traffic-counting system to local governments. In 1975, he dropped out of his law course at Harvard to found the Microsoft Software Company in Washington. Gate's domination of the emerging computer industry began in 1980—1981, when he devised an operating system and licensed it to IBM. MS-DOS became the standard operating system for nearly all IBM personal computers, during the 1980s. Microsoft also developed more specialized software. When the company went public in 1986, Gates became a multimillionaire at the age of 31. Five years later, he was ranked as the world's richest man. In the 1990s, Gates made a fresh fortune from sales of Windows, a system that enables a computer to be operated with on-screen symbols rather than complex keyboard commands. A revised version was launched amid huge publicity in 1995.

Questions 29 to 31 are based on the passage you have just heard.

【内容概要】本文主要介绍了比尔·盖茨的生平。

29. What did Gates do at the age of 14?

【解析】细节题。文中第二句话 At the age of 14, he founded a computer programming company with three friends, and they had earned $20,000 by selling their traffic-counting system to local governments.（14岁的时候，他和三个朋友共同开了一家公司，并且向当地政府出售交通计算系统程序，赢利$20,000。）可知答案是 A) He sold traffic-counting system to local governments.

30. Which of the following had been the standard operating system for nearly all IBM personal computers?

【解析】细节题。从…MS-DOS became the standard operating system for nearly all IBM personal computers 判断答案是 C) MS-DOS.

31. When did Bill Gates become the world's richest man?

【解析】细节题。从…a multimillionaire at the age of 31. Five years later, he was ranked as the world's richest man. 判断是 B) At the age of 36.

Passage Three

For more than six million American children, coming home after school means coming home to an empty house.

135

Some deal with the situation by watching TV. Some may hide. But all of them have something in common. They spend part of each day alone. They are called latchkey children. They're children who look after themselves while their parents work. And their bad condition has become a subject of concern.

A headmaster of an elementary school said that there was a school rule against wearing jewelry. A lot of kids had chains around their necks with keys attached. He was constantly telling them to put them inside their shirts. There were so many keys. Slowly, he learned they were house keys.

He began talking to the children who had them. Then he learned of the impact working couples and single parents were having on their children. Fear is the biggest problem faced by children at home alone. Many had nightmares and were worried about their own safety.

The most common way latchkey children deal with their fears is by hiding. It might be in a bathroom, under a bed or in a closet. The second way is TV. They'll often play it at high volume. Most parents don't realize the effect they have on their children when they leave their children alone.

Questions 32 to 35 are based on the passage you have just heard.

【内容概要】本文主要是讨论由于父母忙于工作，而放学独自一人在家的孩子们的现状，从而也提出一些问题，例如，对于孩子们来说最大的问题就是恐惧。

32．What is the meaning of "latchkey" children?

【解析】推断题。从 "…coming home after school means coming home to an empty house." 和 "They spend part of each day alone." 判断 "latchkey" children 指的是 D) Children who suffer problems from being left alone.

33．What did the headmaster ask the children to do?

【解析】细节题。从 "He was constantly telling them to put them inside their shirts. 他让他们把钥匙放进衬衫里。" 判断答案是 B) Hide the keys in their shirts.

34．How do the children feel when they are at home by themselves?

【解析】细节题。从 "Fear is the biggest problem faced by children at home alone. 恐惧是这些独自在家的孩子们的最大问题" 判断答案是 D) Fear.

35．Which conclusion can we draw from the passage?

【解析】综合分析题。从文中最后一句 "Most parents don't realize the effect they have on their children when they leave their children alone." 觉得多数父母没意识到他们把孩子独自留在家的后果。可知选 D) Some parents don't know the impact on children when they leave them alone.

Section C

If parents bring up a child with the (36)aim of turning the child into a (37)genius, they will cause a (38)disaster. According to several leading (39)educational psychologists, this is one of the biggest mistakes that ambitious parents make. Generally, the child will be only too (40)aware of what the parent expects, and will fail. Unrealistic parental expectations can cause great (41)damage to children. However, if parents are not too unrealistic about what they expect their children to do, but are ambitious in a sensible way, the child may (42)succeed in doing very well—especially if the parents are very (43)supportive of their child. Michael Li is very lucky. (44)He is crazy about music, and his parents help him a lot by taking him to concerts and arranging private piano and violin lessons for him. Although Michael's mother knows very little about music, Michael's father plays the trumpet in a large orchestra. However, he never makes Michael enter music competitions if he is unwilling. Michael's friend, Winston Chen, however, is not so lucky. (45)Because both his parents are successful musicians, they set too high standard for Winston. They want their son to be as successful as they are and so they enter him in every piano competition held. They are very unhappy when he does not win. (46)"When I was your age, I used to win every competition I entered." Winston's father tells him. Winston is always afraid that he will disappoint his parents and now he always seems quiet and unhappy.

Part IV   Reading Comprehension (Reading in Depth)
Section A

【短文大意】如果你拥有假 MBA 文凭，你应该避开俄勒冈州。它是美国严厉追捕出售和使用假文凭的少数几个州之一。在其他地区，对使用假文凭的处理绝大部分取决于雇主，但处罚是很严厉的，包括解雇、职业生涯的毁灭，甚至被揭发为骗子。然而，整个美国，成千上万的刻苦的 MBA 持有者显然认为即使存在着暴露的风险也是值得的，因为在大多数地方并没有为雇主提供简单的区分真假文凭的方法。

在有文凭意识的美国，每年都有成千上万的此类证书被"授予"。一些公司出售伪造的文凭，而其他公司则通过贿赂某个人把伪造的文档插入有效的学校系统，从而出售真正的文凭。

不同于假医疗文凭，一份伪造的 MBA 文凭是不会杀死任何人的。但想获得这样一个文凭又是相当昂贵的。公司花时间和工资结果发现持假文凭的人不称职，那岂不是很糟糕的。但如果他在被抓住之前犯了过失，还将在诉讼上花费一大笔钱。因此如何确保一份 MBA 文凭是通过努力获得的而不是买来的呢？伊利诺斯州大学的教授 George Gollin 做了陈述，描述了他如何追踪各种各样的出售假文凭的学校。将学校的名字输入 Google 中，将弹出一个制造假文凭的学校列表。如果一个即将成为雇员的人从伪造假文凭的学校买了文凭，将学校名字输入搜索引擎将会获得相同的结果。

47.【答案及解析】H) 第一句指出拥有假文凭的人应该避开俄勒冈州，空格所在的句子表达的意思是俄勒冈州对假文凭检查得很严格，因此应选 aggressively。gradually 虽然也能修饰动词，但填在这里，句意不通顺，故选 H）。

48.【答案及解析】O) 空格前的 but 表示后一分句与前一分句形成转折，前一分句指出执行主要依靠雇主，后一句形成转折。而后面的 getting sacked 和 being exposed as a fraud 都是惩罚的一种具体手段，进一步提示此空的意思，因此 O)penalties 为正确答案。

49.【答案及解析】M) 后面的 and 说明此空和 getting sacked 及 being exposed as a fraud 属并列，都是一些惩罚措施，因此选 ruining，表示对某人的职业生涯有不良的影响。hurting 一般是指对人的肉体和精神方面造成的伤害，在这里不合题意。

50.【答案及解析】D) 前一句提到持假文凭的人会受到惩罚，在这里 the risks 为其中之一，所以选揭露 exposure。

51.【答案及解析】G) 此处表达的意思是每年都会有很多这样的文凭被授予，故选 awarded。award 的用法是：award sb. sth. reward 是干扰项，reward 的用法是 reward sb. with sth. 或 reward sb. for doing sth.。

52.【答案及解析】N) 空格处应该填形容词来修饰 school's systems，while 前后应为对比关系，前面是 Some companies sell counterfeit diplomas from real schools，因此此处选 N) valid "有效的"。

53.【答案及解析】F) 根据句意应该选 unqualified，表示花时间和工资结果发现 someone（持假文凭的人）不称职是很糟糕的。

54.【答案及解析】C) 空格前面的 rather than 提示所填词与 earned 并列，但表示相反的意思，因此选 bought，表示 MBA 学位是买来的，而不是通过努力获得的，和文章整篇所讲的假文凭正好相符。

55.【答案及解析】L) 此空应该是名词，前文说在 Google 里输入学校名字，那么显示出来的应该是制造假文凭的学校的列表，因此此处选 list。

56.【答案及解析】I) 根据上下文和句意，此处填输入，把资料等输入电脑的动词应该用 typing，即 I）。

Section B

Passage one

What, besides children, connects mothers around the world and across the sea of time ? It is chicken soup, one prominent American food, expert says.

From Russian villages to Africa and Asia, chicken soup has been the remedy for those weak in body and spirit. Mothers passed their knowledge on to ancient writers of Greece, China and Rome, and even the 12th century philosopher and physician Moses Maimonides extolled (赞美) its virtues.

Among the ancients, Aristotle thought poultry should stand in higher estimation than four-legged animals because the air is less dense than the earth. Chickens got another boost in the Book of Genesis, where it is written that birds and fish were created on the fifth

But according to Mimi Sheraton, who has spentmuch of the past three years exploring the world of chicken soup, much of the reason for chicken's real or imagined curative powers comes from its color.

Her new book, The Whole World Loves Chicken Soup, looks at the beloved and mysterious brew, with dozens of recipes from around the world. "Throughout the ages", she said, "there has been a lot of feeling that white-colored foods are easier to eat for the weak-women and the ill.

In addition, "soups, or anything for that matter eaten with a spoon" are considered "comfort foods", Sheraton said.

"I love soup and love making soup and as I was collecting recipes I began to see this as international dish—it has a universal mystique as something curative, a strength builder",Sheraton said from her New York home.

Her book treats the oldest remedy as if it was brand new.

he National Boiler Council, the trade group representing the chicken industry, reported that 51 percent of the people it surveyed said that they bought chicken because it was

57. Which of the following can be the best title of the passage ?

A) Prominent American Foods

B) Chicken Soup Recipes

C) Chicken Soup, a Universal Cure-all

D) History of the Chicken Soup

58. Since ancient times, the value of chicken soup has been _____.

A) widely acknowledged

B) over-estimated

C) appreciated only by philosophers

D) has been known to mothers day, a day before four-legged animals.

59. Chicken soup earns universal praise because _____.

A) chicken soup has a very long history

B) chicken soup has curative power for those weak in body and spirit

C) poultry usually stands higher than four-legged animals

D) birds were said to be created earlier than four-legged animals

60. According to Sheraton, chicken soup has curative powers mainly for its _____.

A) taste

B) color

C) flavor

D) recipe

61. It can be said from the survey that chicken is _____.

A) a popular food

healthier, 50 percent said it was versatile, 41 percent said it was economical and 46 percent said it was low in fat

B) a main dish

C) cheaper than any other food

D) easy to cook

【短文大意】除了孩子之外，什么能把古今中外的母亲联系在一起？是鸡汤，一种重要的美国食物，专家这样说。

从俄罗斯乡村到非洲和亚洲，鸡汤已成为那些虚弱的人身体和心灵的一剂补药。母亲们把她们对鸡汤的认识传承给古希腊、古中国和古罗马作家，甚至12世纪的哲学家、内科医师 Moses Marmonides 都曾赞美它的优点。

古人之中，亚里斯多德认为家禽应该获得比四条腿的动物更高的评价，因为空气比土稀薄。在《创世纪》中鸡的地位有了进一步提高，书中记录了鸟和鱼是在先于四条腿动物一天的第五天创造的。

但是按照花费了近三年时间研究鸡汤的 Mimi Sheraton 的说法，鸡汤具有真实的或想像的医疗作用的绝大部分原因在于它的颜色。

她的新书 The Whole World Loves Chicken Soup，着眼于运用世界各地的处方而进行了用心的，神秘的熬制。她说，"从古至今，白色食物都容易被虚弱的妇女和病人接受"。

此外，Sheraton 补充说："汤或任何用汤匙吃的食品都会让人感到很舒服"。

"我喜爱汤，喜爱做汤，因此我收集秘方并把鸡汤看作是一道国际美食——它作为具有医疗作用及力量的提供有一种普遍的神秘性"，Sheraton 在她纽约家中这样说。

她的这本书全新地探讨了这一古老的治疗法。

代表鸡产业的行业集团报道，在所调查的购买鸡肉的所有人中，51%的人认为它有助于健康，50%的人认为它有多方面的效用，41%的人认为它经济实惠，46%的人认为它的脂肪含量低。

57. 【题目译文】这篇文章最好的题目是哪一个？

【答案及解析】C) 主旨题。本文并未介绍鸡汤的历史及鸡汤的做法，因此排除 B)和 D)。A)也不正确。本文第二、三段指出，鸡汤受到世界各地的人们的青睐有历史的、宗教的渊源和医疗的作用。故而 C)为正确选项。

58. 【题目译文】从古代开始，鸡汤的价值已经被_____。

【答案及解析】A) 文章第二、三段介绍了历史上人们对鸡汤的青睐，对鸡汤价值的广泛认可，因此 A) 为正确选项。根据 Mimi Sheraton 的观点，鸡汤的医疗作用有些是真实的，有些是想像的，但不能说其价值被过高地估计了，因此 B)不对，C)和 D)过于片面。

59. 【题目译文】鸡汤赢得普遍的赞扬是因为_____。

【答案及解析】B) 第二段第一句话"From Russian villages to Africa and Asia, chicken soup has been the remedy for those weak in body and spirit."即：从俄罗斯乡村到非洲和亚洲，鸡汤已成为那些虚弱的人身体和心灵的一剂补药。由此可以推断 B)为正确选项。

60. 【题目译文】通过 Sheraton，鸡汤具有医疗作用主要是因为_____。

【答案及解析】B) 在文章第四段，Mimi Shcraton 认为鸡汤的真实或想像的医疗作用大多是它的颜色的原因。因此 B)为正确选项。

61. 【题目译文】从调查可以看出鸡是_____。

【答案及解析】A) 文章最后一段介绍了一项调查，结果表明，大多数人喜欢鸡汤是因为他们认为鸡汤是有益健康的食品，因此 A)为正确选项。

Passage Two

Most episodes of absent-mindedness—forgetting where you left something or wondering why you just entered a room—are caused by a simple lack of attention, says Schacter. "You're supposed to remember something, but you haven't encoded it deeply."

Encoding, Schacter explains, is a special way of paying attention to an event that has a major impact on recalling it later. Failure to encode properly can create annoying situations. If you put your mobile phone in a pocket, for example, and don't pay attention to what you did because you're involved in a conversation, you'll probably forget that the phone is in the jacket now hanging in your wardrobe . "Your memory itself isn't failing you," says Schacter. "Rather, you didn't give your memory system the information it needed."

Lack of interest can also lead to absent-mindedness. "A man who can recite sports statistics from 30 years ago," says Zelinski, "may not remember to drop a letter in the mailbox."

Women gave slightly better memories than men, possibly because they pay more attention to their environment, and memory relies on just that.

Visual cues can help prevent absent-mindedness, says Schacter. "But be sure the cue is clear and available," he cautions. If you want to remember to take a medication (药物)with lunch, put the pill bottle on the kitchen table—don't leave it in the medicine chest and write yourself a note that you keep in a pocket.

Another common episode of absent-mindedness: walking into a room and wondering why you're there. Most likely, you were thinking about something else. "Everyone does this from time to time," says Zelinski. The best thing to do is to return to where you were before entering the room. And you'll likely remember.

62. Why does the author think that encoding properly is very important ?

A) It helps us understand our memory system better

B) It enables us to recall something from our memory

C) It expands our memory capacity considerably

D)It slows down the process of losing our Memory

63. One possible reason why women have better memories than men is that _____ .

A) they have a wider range of interests

B) they are more reliant on the environment

C) they have an unusual power of focusing their attention

D) they are more interested in what's happening around them

64. A note in the pocket can hardly serve as a reminder because _____ :

A) it will easily get lost

B) it's not clear enough for you to read

C) it's out of your sight

D) it might get mixed up with other things

65. What do we learn from the last paragraph?

A) If we focus our attention on one thing, we might forget another.

B) Memory depends to a certain extent on the environment.

C) Repetition helps improve our memory.

D) If we keep forgetting things, we'd better return to where we were.

66. What is the passage mainly about ?

A) The process of gradual memory loss.

B) The causes of absent-mindedness. (主旨大意)

C) The impact of the environment on memory.

D) A way of encoding and recalling.

【短文大意】许多健忘的生活小插曲——忘记东西放在哪儿或不知为什么进房间——仅仅是由于没有用心的缘故。Schacter 说："你本应该能记住一件事情，但是却没有把它进行编码从而深深地印在你的大脑里。"

140

Schacter 解释说，编码就是注意一件事情的特殊方式，它会对你将来是否能够回忆起这件事有影响。不能合理地编码让人非常烦恼。例如，你在和别人谈话时不自觉地把手机放进衣袋里，那么你就会很可能记不得其实你的手机就在衣柜的夹克衫里。Schacter 说，"你的记忆本身没有问题，而是你没有将记忆系统所需的信息传递过去。"

缺少兴趣也会让人健忘。Zelinski 说："一个人可以背出 30 年前体育的统计数据，却很可能忘记自己在邮箱中投过的一封信。"男性在记忆方面稍逊于女性，原因很可能是女性更留意身边的环境，而记忆正是依赖于此。

Schacter 说，视觉提示有助于消除健忘。他告诫说，"但你得确保这些提示清晰可见。"如果你想记住午餐时服药，那么就将药瓶放进厨房的桌子上——不要把药瓶放在装药的抽屉里，而兜里一直揣着提示你吃药的纸条。

另外一个健忘的常见的插曲就是：当你走进一间屋子时，却不知到来这是干什么。这很可能是因为你在想其他事情。Zelinski 说，"每个人偶尔都会有这样的事情发生。要想避免，最好的做法是回到进房间之前呆的地方，这样你可能就会想起你要做的事情。"

62. 【题目译文】为什么作者认为正确编码非常重要？

【答案及解析】B) 见第二段第一句话，"Encoding, Schacter explains, is a special way of paying attention to an event that has a major impact on recalling it later." 即：编码就是注意一些事情的特殊方式，它对你将来能把这件事情回忆起来会产生重大影响。由此可以确定 B)为正确选项。

63. 【题目译文】女性记忆比男性记忆好的原因可能是_____。

【答案及解析】D) 见第三段最后一句话，"Women gave slightly better memories than men, possibly because they pay more attention to their environment, and memory relies on just that."句中"that"指的就是上一句所说的"pay more attention to their environment"。女性记忆比男性记忆好的原因可能是女性更留意身边的环境，而记忆正是依赖于此。因此 D)为正确选项。

64. 【题目译文】放在衣兜里的纸条几乎发挥不了提示作用的原因是_____。

【答案及解析】C) 因为放在衣兜里的纸条不在人的视线之内，因此它发挥不了提示作用。由文章第四段可以推断出 C)为正确选项。

65. 【题目译文】从最后一段我们知道_____。

【答案及解析】A) 见文章最后一段的举例：当你走进一间屋子时，却不知道来这是干什么。这很可能是因为你在想其他事情。由此可以推断出 A)为正确选项。

66. 【题目译文】本文主要讲的是什么？

【答案及解析】B) 主旨题。B)为正确选项，因为全文都在讲述和分析健忘的原因。

Part V Cloze

【短文大意】在今天，对于许多人来说，阅读不再是一份休闲。为了跟上工作进程，他们必须阅读信件、报告、交易文件、办公室内部信息，更不用说报纸和杂志了：这是一份没有止境的文字工作。为了得到一份工作或者提高自己，快速阅读和理解能力是关系到成败的关键所在。然而，很不幸，大多数人都是阅读慢的人。我们当中的大多数人早期都养成了不好的阅读习惯，并且从来不克服这一点。主要的缺陷在于语言的自身要素，即"单词"。如果单个地看这些字，它们并没有什么意义，但是，如果将它们组合成词组、句子和段落，它们则有了意义。从逻辑上来说，未受过阅读训练的人不会理解意群。他费力地一次读一个单词，经常回顾刚才看过的单词或段落。返回，这种回头看刚才所阅读过的内容的行为趋势是阅读中的一个最普遍的坏习惯。另一个减慢阅读速度的习惯是"发音"——在阅读者阅读时，在口头或者在心里去读出每一个单词。

为了克服这些坏习惯，一些阅读讲义团体使用一个叫做加速器的设备，这种设备以预先确定的速度把当前页遮住。这个器械被设置的速度比阅读者之前所适应的速度要快，用来促进阅读者。快速阅读器迫使阅读者加快阅读速度，使阅读者再也不能逐字阅读、回顾前文内容或者默读。速度最初会影响理解，但是当学着阅读文章大意和概念时，就会发现，不仅能够读得更快，而且理解能力也得以提高。许多人已经发现他们的阅读技巧在经过训练后已经明显地得到了提高。以 Charlce Au 为例，他是一个商业管理者，他的阅读速度在培训前是一分钟 172 个单词，现在他已经达到了一个非常好的速度，每分钟 1,378 个单词。现在他非常高兴能够在较短时间内读完众多的材料。

67.【答案】D 本句意思是"如果谁想谋得一份差事"。applying 需加 for,意思是"申请"；B.doing 做；C.offering 提供，此三项均不符题意，只有 D.getting（获得）适合。

68.【答案】A 本句意为"快速阅读与理解的能力，是关系到成败的关键所在"只有 quickly 与原意吻合。easily（容易地）、roughly（粗略地）、decidedly（果断地）均与原文内容不符。

69.【答案】C 英语中，阅读速度快的人称为 good reader，反之，就是 poor reader。根据上下文的内容，多数人都属于 poor reader，因此选 poor（差的）。其他选项不妥。

70.【答案】B 此处的意思是"大多数人早期养成看书慢的习惯"，因此选 habits（习惯）。training 训练，培训；situations 形势；custom 风俗习惯。

71.【答案】A 此处说的是"主要的困难在于语言的自身要素，即单词"。combines 联合；touches 接触；involves 包括，这三项的词义与原文不符。而 lies 与 in 构成搭配，意为"在于"，所以选 A。

72.【答案】C 这里的意思是"如果单个地看这些字，它们并没有什么意义"。此处需要一个否定意义的单词。some 有点；a lot 许多；dull 单调的。这三项均不合题意。只有 little（很少）是否定词，合乎逻辑，所以 C 正确。

73.【答案】D 此句意为"作者对未受过阅读训练的人的不良习惯感到遗憾"。Fortunately（幸运地）；In fact 事实上；Logically（合乎逻辑地）均不妥，Unfortunately（不幸地）合乎句义。

74.【答案】B 此句意为"在阅读时经常重读（反复读）"因此，选 B reread（重读）。reuse 再使用；rewrite 改写；recite 背诵。

75.【答案】A 此处所填的词既是 look back over 的宾语，又是 you have just read 的宾语，只有 what 能充当这种双重成分，所以选 A。

76.【答案】C scales down 按比例减少；cuts down 削减，此两项不合题意。measures 不能与 down 搭配。只有 slow 与 down 搭配的意思"放慢"在此合适，所以选 C。

77.【答案】B 本段前文已经出现 you，在此选 one（泛指人们，我们，你）来代替 you；some one 无此用法；如果用 reader，前面应加定冠词；he 不能与该段逻辑一致。

78.【答案】A 此句意为"训练快速阅读所使用的工具必然与提高阅读速度有关"，因此选 accelerator（快读器）。actor 演员；amplifier 放大器；observer 观察者。

79.【答案】D 前面的 faster 决定了应当选 than，构成比较级。

80.【答案】C 此句意为"快速阅读器迫使你加快阅读速度，使你再也不能逐字阅读，回顾前文内容或者默读"。enabling 相当于 making possible；leading 引导；indicating 指出，表明，这三项都不合题意。只有 making（使，使得）最合适。

81.【答案】B 这里的意思是"速读最初会影响理解"，所以选 comprehension（理解）。meaning 意义，意思，指词或词组表示的意义；gist 大意，要旨；regression 回顾。

82.【答案】A 与前半句中的 not only 相呼应，构成句式"不仅……，而且……"，只有选 but，而 nor、or 或 for

均不能构成固定用法。

83.【答案】C 本句中的主语是第三人称复数，物主代词必然是 their，所以选 C。

84.【答案】B take 与后面的 for instance 构成短语，意为："以……例"，其他三项不能构成搭配。

85.【答案】D 这里提到将受训之前与受训之后进行对比，因此选 D before，表示受训之前，其他三项均不合题意。

86.【答案】D 此处意为：在较短时间内读完众多的材料。master 掌握；go over 复习；present 呈现，展现，此三项均不妥，只有 get through （读完）最恰当。

Part VI   Translation

87．My mother would _____(大发脾气) if she found out about this.

【题目译文】如果我的妈妈发现了这件事，她会大发脾气的。

【答案及解析】have a fit 注意 "have a fit" 是口语形式，表达的意思为 "大吃一惊，大发脾气"。

88．Nobody knows how the universe _____(开始形成) .

【题目译文】没有人知道宇宙是怎样开始形成的。

【答案及解析】came into being come into being：（事物）产生，出现。

89．Furniture, books, clothes——_____(所有能够说得出来的) they sell it.

【题目译文】他卖掉了所有能够说得出来的东西：家具、书籍、衣服。

【答案及解析】you name it 注意 "it" 是指代 "Furniture, books, clothes" 这一整体概念的。

90．Let's start again from where we _____(停止) .

【题目译文】让我们从停止的地方再次开始。

【答案及解析】left off leave off：（使）停止，（使）结束。

91．What you said has_____(使人注意) to that matter.

【题目译文】你所说的话已经使人注意到那件事了。

【答案及解析】drawn someone's attention   draw one's attention：吸引某人的

《大学英语4级考试全真模拟试卷及详解（修订本）》
意见反馈表

    为了进一步改进我们的工作，使该书更具人性化，编排设计更合理，更好地服务于考生，望您能将宝贵的意见填在下表并寄到我中心。

您认为8套模拟题题量合适吗？
详解部分您认为有哪些还需要改进？
其他意见和建议
您的姓名、联系地址、邮编、电话

回信请寄： 北京市安定门外安华里二区一号楼   石油工业出版社

          社会图书出版中心   邓晓素 收   邮政编码：100011

电子信箱：dengxiaosu_201@sina.com（复印有效）

图书在版编目（CIP）数据

大学英语 4 级考试 710 分全真模拟试卷及详解(修订本)/ 王波等主编.
北京：石油工业出版社，2008.10
（江涛英语）
ISBN 978 – 7 – 5021 – 6794 – 3

Ⅰ. 大…
Ⅱ. 王…
Ⅲ. 英语–高等学校–水平考试–解题
Ⅳ. H319.6

中国版本图书馆 CIP 数据核字（2008）第 154074 号

大学英语 4 级考试 710 分全真模拟试卷及详解（修订本）

王波　张琳　周天楠 主编

出版发行：石油工业出版社

（北京安定门外安华里 2 区 1 号　100011）

网　址：www.petropub.com.cn

发行部：(010) 64523604　　　　编辑部：(010) 64523615

经　销：全国新华书店

印　刷：北京市朝阳燕华印刷厂

2008 年 10 月第 2 版　2008 年 10 月第 4 次印刷

787×1092 毫米　开本：1/16　印张：18.00

字数：500 千字

定价：26.80 元

（如出现印装质量问题，我社社会图书出版中心营销部负责调换）

Model Test One

Part I   **Writing**   **(30 minutes)**

注意：此部分试题在**答题卡 1** 上

Part Ⅱ   **Reading Comprehension (Skimming and Scanning)**   **(15 minutes)**

Directions: *In this part, you will have 15 minutes to go over the passage quickly and answer the questions on **Answer Sheet 1**.*

For questions 1～7, mark

    Y (for YES)   *if the statement agrees with the information given in the passage;*

    N (for NO)   *if the statement contradicts the information given in the passage;*

    NG (for NOT GIVEN)   *if the information is not given in the passage.*

For questions 8～10, complete the sentence with the information given in the passage.

注意：此部分试题请在**答题卡 1** 上作答

Wildfires

In just seconds, a spark or even the sun's heat alone sets off an extremely large fire. The wildfire quickly spreads, consuming the thick, dried-out plants and almost everything else in its path. What was once a forest becomes a virtual powder keg of untapped fuel? In a seemingly instantaneous burst, the wildfire overtakes thousands of acres of surrounding land, threatening the homes and lives of many in the vicinity.

Fire Starts

On a hot summer day, when drought conditions peak, something as small as a spark from a train car's wheel striking the track can ignite a raging wildfire. Sometimes, fires occur naturally, ignited by heat from the sun or a lightening strike. However, the majority of wildfires are the result of human carelessness.

Common causes for wildfires include:

Arson

Campfires

Discarding lit cigarettes

Improperly burning debris

Playing with matches or fireworks

Prescribed fires

Everything has a temperature at which it will burst into flames. This temperature is called a material's flash point. Wood's flash point is 572 degrees Fahrenheit (300℃). When wood is heated to this temperature, it releases hydrocarbon gases that mix with oxygen in the air, combust and create fire.

There are three components needed for ignition and combustion to occur. A fire requires fuel to burn, air to supply oxygen, and a heat source to bring the fuel up to ignition temperature. Heat, oxygen and fuel form the fire triangle. Firefighters often talk about the fire triangle when they are trying to put out a blaze. The idea is that if they can take away any one of the pillars of the triangle, they can control and ultimately extinguish the fire.

After combustion occurs and a fire begins to burn, there are several factors that determine how the fire spreads. These three factors include fuel, weather and topography. Depending on these factors, a fire can quickly fizzle or turn into a raging blaze that scorches thousands of acres.

Fueling the Flames

Wildfires spread based on the type and quantity of fuel that surrounds it. Fuel can include everything from trees, underbrush and dry grassy field to homes. The amount of flammable material that surrounds a fire is referred to as the fuel load. Fuel load is measured by the amount of available fuel per unit area, usually tons per acre.

A small fuel load will cause a fire to burn and spread slowly, with a low intensity. If there are a lot of fuels, the fire will burn more intensely, causing it to spread faster. The faster it heats up the material around it, the faster those materials can ignite. The dryness of the fuel can also affect the behavior of the fire. When the fuel is very dry, it is consumed much faster and creates a fire that is much more difficult to contain.

Here are the basic fuel characteristics that decide how it affects a fire.

Size and shape

Arrangement

Moisture content

Small fuel materials, also called flashy fuels, such as dry grass, pine needles, dry leaves, twigs and other dead brush, burn faster than large logs or stumps (this is why you start a fire with kindling rather than logs). On a chemical level, different fuel materials take longer to ignite than others. But in a wildfire, where most of the fuel is made of the same sort of material, the main variable in ignition time is the ratio of the fuel's total surface area to its volume. Since a twig's surface area is much larger than its volume, it ignites quickly. By comparison, a tree's surface area is much smaller than its volume, so it needs more time to heat up before it ignites.

As the fire progresses, it dries out the material just beyond it—heat and smoke approaching potential fuel causes the fuel's moisture to evaporate. This makes the fuel easier to ignite when the fire finally reaches it. Fuels that are somewhat spaced out will also dry out faster than fuels that are packed tightly together, because more oxygen is available to the thinned-out fuel. More tightly-packed fuels also retain more moisture, which absorbs the fire's heat.

Wind and Rain

Weather plays a major role in the birth, growth and death of a wildfire. Drought leads to extremely favorable conditions for wildfires, and winds aid a wildfire's progress—weather can spur the fire to move faster and engulf more land. It can also make the job of fighting the fire even more difficult.

There are three weather ingredients that can affect wildfires:

Temperature

Wind

Moisture

As mentioned before, temperature has a direct effect on the sparking of wildfires, because heat is one of the three pillars of the fire triangle. The sticks, trees and underbrush on the ground receive radiant heat from the sun, which heats and dries potential fuels. Warmer temperature allow for fuels to ignite and burn faster, adding to the rate at which a wildfire spreads. For this reason, wildfires tend to rage in the afternoon, when temperatures are at their hottest.

Wind probably has the biggest impact on a wildfire's behavior. It's also the most unpredictable factor. Winds supply the fire with additional oxygen, further dry potential fuel and push the fire across the land at a faster rate.

The stronger the wind blows, the faster the fire spreads. The fire generates wind of its own that are as many as 10 times faster than the wind of surrounding area. It can even throw embers into the air and create additional fires, an occurrence called spotting. Wind can also change the direction of the fire, and gusts can raise the fire into the trees, creating a crown fire.

While wind can help the fire to spread, moisture works against the fire. Moisture, in the form of humidity and precipitation, can slow the fire down and reduce its intensity. Potential fuels can be hard to ignite if they have high levels of moisture, because the moisture absorbs the fire's heat. When the humidity is low, meaning that there is a low amount of water vapor in the air, wildfires are more likely to start. The higher the humidity, the less likely the fuel is to dry and ignite.

Since moisture can lower the chances of a wildfire igniting, precipitation has a direct impact on fire prevention. When the air becomes saturates with moisture, it releases the moisture in the form of rain. Rain and other precipitation raise the amount of moisture is fuel, which suppresses any potential wildfires from breaking out.

Fire on the Mountain

The third big influence on wildfire behavior is the lay of the land, or topography. Although it remains virtually unchanged, unlike fuel and weather, topography can either aid or hinder wildfire progression. The most important factor in topography as it relates to wildfire is slope.

Unlike humans, fires usually travel uphill much faster than downhill. The steeper the slope, the faster the fire travels. Fires travel in the direction of the ambient wind, which usually flows uphill. Additionally, the fire is able to preheat the fuel further up the hill because the smoke and heat are rising in that direction. Conversely, once the fire has reached the top of a hill, it must struggle to come back down because it is not able to preheat the downhill fuel as well as the uphill.

In addition to the damage that fires cause as they burn, they can also leave behind disastrous problems, the effects of which might not be felt for months after the fire burns out. When fires destroy all the vegetation on a hill or mountain, it can also weaken the organic material in the soil and prevent water from penetrating the soil. One problem that results from this is extremely dangerous erosion that can lead to debris flows.

While we often look at wildfires as being destructive, many wildfires are actually beneficial. Some wildfires burn the underbrush of a forest, which can prevent a larger fire that might result if the brush were allowed to accumulate for a long time. Wildfires can also benefit plant growth by reducing disease, releasing nutrients from burned plants into the ground and encouraging new growth.

Questions:

1. This passage explores how wildfires are born and live.

2. Wood will burst into flames at 300 degrees Fahrenheit.

3. Wildfires are mainly caused by people, not by nature.

4. In most situations, we measure fuel load by tons per hectare.

5. The wind generated by the fire itself are a little slower than the wind of the surrounding area.

6. Fires are like humans in that they usually travel downhill much faster than uphill.

7. In fact, the effects of fires might not be felt for months.

8～10 题请见答题卡 1

Part Ⅲ   Listening Comprehension   (35 minutes)

Section A

Directions: *In this section, you will hear 8 short and 2 long conversations. At the end of each conversation, one or more questions will be asked about what was said. Both the conversation and the question will be spoken only once. After each question there will be a pause. During the pause, you must read the four choices marked A), B), C) and D), and decide which is the best answer. Then mark the corresponding letter on the **Answer Sheet 2** with a single line through the center.*

注意：此部分试题请在**答题卡 2** 上作答

11. A) Twenty-five dollars.   B) Twenty dollars.
     C) Forty dollars.   D) Fifty dollars.

12. A) To go to the French restaurant.   B) To try a new restaurant.
     C) To visit a friend.   D) To stay at home.

13. A) Easy-going and friendly.   B) Very nervous.
     C) Angry.   D) Not easy-going.

14. A) He plays jazz music.   B) He is a jazz fan.
     C) He needs 300 jazz records.   D) He likes classical music.

15. A) At a post office.   B) At a bank.
     C) At a restaurant.   D) At an airport.

16. A) He was scared.
    C) He hasn't got a car.
    B) He was upset.
    D) He is glad to drive her there.
17. A) Lending money to a student.
    C) Reading a student's application.
    B) Filling a form.
    D) Asking for some financial aid.
18. A) 12:30   B) 11:30   C) 12:00   D) 11:00

Questions 19 to 22 are based on the conversation you have just heard.

19. A) The benefits of strong business competition.
    B) A proposal to lower the cost of production.
    C) Complaints about the expense of modernization.
    D) Suggestions concerning new business strategies.

20. A) It costs much more than its worth.
    B) It should be bought up-to-date.
    C) It calls for immediate repairs.
    D) It can still be used for a long time.

21. A) The personnel manager should be fired for inefficiency.
    B) A few engineers should be employed to modernize the factory.
    C) the entire staff should be retrained.
    D) Better-educated employees should be promoted.

22. A) Their competitors have long been advertising on TV.
    B) TV commercials are less expensive.
    C) Advertising in newspaper alone is not sufficient.
    D) TV commercials attract more investments.

Questions 23 to 25 are based on the conversation you have just heard.

23. A) Searching for reference material.
    C) Writing a course book.
    B) Watching a film of the 1930s'.
    D) Looking for a job in a movie studio.

24. A) It's too broad to cope with.
    C) It's controversial.
    B) It's a bit outdated.
    D) It's of little practical value.

25. A) At the end of the online catalogue.
    B) At the Reference Desk.
    C) In the New York Times.
    D) In the Reader's Guide to Periodical Literature.

Section B

Directions: *In this section, you will hear 3 short passages. At the end of each passage, you will hear some questions. Both the passage and the questions will be spoken only once. After you hear a question, you must choose the best answer from the four choices marked A), B), C) and D). Then mark the corresponding letter on the **Answer Sheet 2** with a single line* *through the center.*

注意：此部分试题请在**答题卡 2** 上作答

Passage One

Questions 26 to 28 are based on the passage you have just heard.

26. A) Americans are too attached to their cars.

    B) American cars are too fast.

    C) Automobiles endanger health.

    D) Automobiles are the main public transportation tools in USA.

27. A) Because they pollute air.

    B) Because they are natural hazards.

    C) Because they are increasing in numbers.

    D) Because people don't walk so often.

28. A) Control of natural hazards.   B) Control of heavy traffic.

    C) Control of heart disease.   D) Control of man-made hazards.

Passage Two

Questions 29 to 32 are based on the passage you have just heard.

29. A) It smashed into a row of houses.

    B) It was run over by a truck.

    C) It was too nervous to leave the strip in the middle of the road.

    D) It hit a truck.

30. A) A passenger.   B) The dog.

    C) The truck driver.   D) A policeman.

31. A) In the street.   B) In a family swimming pool.

    C) In a public swimming pool.   D) In a kindergarten.

32. A) A big steak.   B) A piece of bread.

    C) A bottle of milk.   D) An extra bone.

Passage Three

Questions 33 to 35 are based on the passage you have just heard.

33. A) Seasonal variations in nature.

    B) How intelligence changes with the change of seasons.

    C) How we can improve our intelligence.

    D) Why summer is the best season for vacation.

34. A) Summer.   B) Winter.

    C) Fall.   D) Spring.

35. A) All people are less intelligent in summer than in the other seasons of the year.

    B) Heat has no effect on people's mental abilities.

    C) People living near the equator are the most intelligent.

    D) Both climate and temperature exert impact on people's intelligence.

Section C

Directions: *In this section, you will hear a passage three times. When the passage is read for the first time, you should listen carefully for its general idea. When the passage is read for the second time, you are required to fill in the blanks numbered from 36 to 43 with the exact words you have just heard. For blanks numbered from 44 to 46 you are required to fill in the missing information. You can either use the exact words you have just heard or write down the main points in your own words. Finally, when the passage is read for the third time, you should check what you have written.*

注意：此部分试题在**答题卡 2 上**，请在**答题卡 2 上作答**

Part IV   Reading Comprehension (Reading in Depth)   (25 minutes)

Section A

Directions： *In this section, there is a passage with ten blanks. You are required to select one word for each blank from a list of choices given in a word bank following the passage. Read the passage through carefully before making your choices. Each choice in the bank is identified by a letter. Please mark the corresponding letter for each item on **Answer Sheet 2** with a single line through the centre. **You may not use any of the words in the bank more than once.***

Questions 47 to 56 are based on the following passage.

The first modern Olympic Games was held in Athens in 1896 and only twelve nations participated. Besides the host nation man participants were tourists who __47__ to be in Greece at the time. Though the whole affair was __48__ and the standard was not high, the old principle of amateur sport was kept up.

Since then the games had been held every four years except during the __49__ of the two World Wars. This was __50__ a departure from the old Olympic spirit when wars had to stop and make way for the games.

The games have grown enormously in scale and __51__ performances have now reached unprecedented heights. Unfortunately the same cannot be said about their __52__ standard. Instead of Olympia, the modern games are now held in different cities all over the world. Inevitably politics and commercialism get involved as countries vie each other for the __53__ to hold the games because of the political prestige and commercial profit to be __54__ out of them. In the 11th games held in Berlin in 1936, Hitler who had newly come to __55__ in Germany tried to use the occasion for his Nazi propaganda. For the first time the Olympic flame was brought all the way from Olympia to the games site in relays, a marathon journey now often taking months to __56__ .

注意：此部分试题请在**答题卡 2 上作答**

A)	honour	I)	happened
B)	accomplish	J)	definitely
C)	had	K)	physical
D)	moral	L)	informal
E)	arrive	M)	interruption
F)	occurred	N)	especially
G)	end	O)	irregular
H)	power		

Section B

Directions: *There are 2 passages in this section. Each passage is followed by some questions or unfinished statements. For each of them there are four choices marked A), B), C) and D). You should decide on the best choice and mark the corresponding letter on the Answer Sheet 2 with a single line through the centre.*

Passage one

Questions 57 to 61 are based on the following passage.

A subject which seems to have been insufficiently studied by doctors and psychologists is the influence of geography and climate on the psychological health of humankind. There seems no doubt that the general character of landscape, the relative length of day and night, and climate must all play a part in determining what kind of people we are.

It is true that a few studies have been made. Where all the inhabitants of a particular area enjoy exceptionally good or bad health, scientists have identified contributory factors such as the presence or absence of substances like iodine, fluoride, calcium, or iron in the water supply, or perhaps types of land that provide breeding places for pests like mosquitoes or rats.

Moreover, we can all generalize about types of people we have met. Those living in countries with long dark winters are apt to be less talkative and less lively than inhabitants of countries where the climate is more equable. And where olives and oranges grow, the inhabitants are cheerful, talkative, and casual.

But these commonplace generalizations are inadequate—the influence of climate and geography should be studied in depth. Do all mountain dwellers live to a ripe old age? Does the drinking of wine, rather than beer, result in a sunny and open character? Is the strength and height of a Kenyan tribe due to their habitual drinking of the cow blood?

We are not yet sure of the answer to such questions, but let us hope that something beneficial to humankind may eventually result from such studies.

注意：此部分试题请在答题卡 2 上作答

57. The author's purpose of writing this passage is to _____.

   A) alert readers to the scarcity of natural resources

B) call for more research on the influence of geographical environment

C) introduce different elements in character cultivation

D) draw more attention to the health condition of mankind

58. It can be inferred that proper amounts of iodine, fluoride and calcium can _____.

A) benefit people's physical health

B) influence the quality of water supply

C) help provide breeding places for pests

D) strengthen a person's character

59. How does the author evaluate the generalizations of people's types in Para. 3?

A) Such generalizations help us judge the different characters of people we meet.

B) Such generalizations are not inclusive enough to draw a convincing conclusion.

C) Such generalizations prove that nature plays an important role in determining social habits.

D) Such generalizations show that there are mainly two different types of people on the planet.

60. According to the passage, research into the influence of climate and geography should _____.

A) focus on unknown aspects   B) be pursued on a larger scale

C) be carried out among remote tribes   D) go ahead in depth

61. What do we know about the generalizations of people's type?

A) People who like drinking wine tend to be optimistic.

B) People who live in mountain areas tend to have a long life.

C) People who live in areas with stable climate tend to be talkative and lively.

D) People who like drinking cow blood tend to be strong and tall.

Passage Two

Questions 62 to 66 are based on the following passage.

To forgive may be divine, but no one said it was easy. When someone has deeply hurt you, it can be extremely difficult to let go of your complaints and hatred. But forgiveness is possible— and it can be surprisingly beneficial to your physical and mental health.

"People who forgive show less depression, anger and stress and more hopefulness," says Frederic, Ph.D., author of *Forgive for Good*. "So it can help save on the wear and tear on our organs, reduce the wearing out of the immune system and allow people to feel more vital."

So how do you start the healing? Try following these steps:

Calm yourself. To remove your anger, try a simple stress-management technique, "Take a couple of breaths and think of something that gives you pleasure: a beautiful scene in nature, someone you love," Frederic says.

Don't wait for an apology. "Many times the person who hurts you has no intention of apologizing," Frederic says. "They may have wanted to hurt you or they just don't see things the same way. So if you wait for people to apologize, you could be waiting an awfully long time." Keep in mind that forgiveness does not necessarily mean reconciliation with the person who upset you or neglecting his or her action.

Take the control away from your offender. Mentally replaying your hurt gives power to the person who caused you pain. "Instead of focusing on your wounded feelings, learn to look for the love, beauty and kindness around you," Frederic says.

Try to see things from the other person's perspective. If you empathize with that person, you may realize that he or she was acting out of ignorance, fear—even love. To gain perspective, you may want to write a letter to yourself from your offender's point of view.

Recognize the benefits of forgiveness. Research has shown that people who forgive report more energy, better appetite and better sleep patterns.

Don't forget to forgive yourself. "For some people, forgiving themselves is the biggest challenge," Frederic says. "But it can rob you of your self-confidence if you don't do it."

注意：此部分试题请在答题卡 2 上作答

62. By saying that forgiveness "can help save on the wear and tear on our organs," Frederic, Ph. D means that _____.

    A) people are likely worn out by crying when they get hurt

    B) we may get physically damaged if we stick to the hurt

    C) our physical conditions benefit most from forgiveness

    D) the immune system is closely related with our organs

63. When you try to calm yourself, you are actually trying to_____.

    A) recall things you love   B) show you are angry

    C) relieve your stress   D) breathe normally

64. Your offender may not want to apologize because _____.

    A) they are afraid that they won't be forgiven

    B) they don't even realize they have hurt you

    C) they don't share the same feeling with you

    D) they think that time can heal any wound

65. You will still be under the control of the offender if _____.

    A) the offender refuses to reconcile with you

    B) you keep reminding yourself on the pain

    C) the offender never feels sorry to you

    D) you don't find love, beauty or kindness

66. Which of the following enables you to gain the offender's perspective?

    A) Empathizing with the offender.

    B) Realizing the reason for the offender's action.

    C) Writing a letter to the offender.

    D) Doing the same thing the offender did to you.

Part V   Cloze   (15 minutes)

Directions: *There are 20 blanks in the following passage. For each blank there are four choices marked A), B), C), and D) on the right side of the paper. You should choose the ONE that best fits into the passage. Then mark the corresponding letter on **Answer Sheet 2** with a single line through the center.*

注意：此部分试题请在**答题卡 2** 上作答

Most people have no idea of the hard work and worry about going into collection of those fascinating birds and animals that they pay to see in the zoo. One of the questions that is always asked of me is ___67___ I became an animal

collector in the first ___68___ .The answer is that I have always been interested in animals and zoos. According to my parents, the first word I was able to say with any ___69___ was not the

conventional "mamma" or "daddy", ___70___ the

word "zoo", which I would ___71___ over and

over again with a shrill ___72___ until someone, in

order to ___73___ me up, would take me to the zoo. When I grow a little older, we lived in Greece and I had a great ___74___ of pets,

___75___ from owls to seahorses, and I spent all my

spare time ___76___ the countryside in search of

fresh specimens to ___77___ to my collection of

pets. ___78___ on I went for a year to the City

Zoo, as a student ___79___ , to get experience of the large animals, such as lions, bears and

67. A. how   B. where
     C. when   D. whether
68. A. region   B. field
     C. place   D. case
69. A. clarity   B. emotion
     C. sentiment   D. affection
70. A. except   B. but
     C. except for   D. but for
71. A. recite   B. recognize
     C. read   D. repeat
72. A. volume   B. noise
     C. voice   D. pitch
73. A. close   B. shut
     C. stop   D. comfort
74. A. many   B. amount
     C. number   D. supply
75. A. changing   B. ranging
     C. varying   D. altering
76. A. living   B. cultivating
     C. reclaiming   D. exploring
77. A. increase   B. include
     C. add   D. enrich
78. A. Later   B. Further
     C. Then   D. Subsequently
79. A. attendant   B. keeper
     C. member   D. assistant

ostriches, __80__ were not easy to keep at home.

When I left, I __81__ had enough money of my

own to be able to __82__ my trip and I have been

going __83__ ever since then. Though a
collector's job is not an easy one and is full of

__84__, it is certainly a job which will appeal

__85__ all those who love animals and __86__.

80. A. that   B. they
    C. of which   D. which
81. A. luckily   B. barely
    C. nearly   D. successfully
82. A. pay   B. provide
    C. allow   D. finance
83. A. normally   B. regularly
    C. usually   D. rarely
84. A. expectations B. sorrows
    C. excitement D. disappointments
85. A. for   B. with
    C. to   D. from
86. A. excursion   B. travel
    C. journey   D. trip

Part VI   Translation   (5 minutes)

Directions: *Complete the sentences on Answer Sheet 2 by translating the Chinese given in brackets into English.*

注意：此部分试题在**答题卡 2** 上，请在**答题卡 2** 上作答

答题卡 1 （Answer Sheet 1）

准　考　证　号

[0]	[0]	[0]	[0]	[0]	[0]	[0]	[0]	[0]	[0]	[0]	[0]	[0]	[0]	[0]
[1]	[1]	[1]	[1]	[1]	[1]	[1]	[1]	[1]	[1]	[1]	[1]	[1]	[1]	[1]
[2]	[2]	[2]	[2]	[2]	[2]	[2]	[2]	[2]	[2]	[2]	[2]	[2]	[2]	[2]
[3]	[3]	[3]	[3]	[3]	[3]	[3]	[3]	[3]	[3]	[3]	[3]	[3]	[3]	[3]
[4]	[4]	[4]	[4]	[4]	[4]	[4]	[4]	[4]	[4]	[4]	[4]	[4]	[4]	[4]
[5]	[5]	[5]	[5]	[5]	[5]	[5]	[5]	[5]	[5]	[5]	[5]	[5]	[5]	[5]
[6]	[6]	[6]	[6]	[6]	[6]	[6]	[6]	[6]	[6]	[6]	[6]	[6]	[6]	[6]
[7]	[7]	[7]	[7]	[7]	[7]	[7]	[7]	[7]	[7]	[7]	[7]	[7]	[7]	[7]
[8]	[8]	[8]	[8]	[8]	[8]	[8]	[8]	[8]	[8]	[8]	[8]	[8]	[8]	[8]
[9]	[9]	[9]	[9]	[9]	[9]	[9]	[9]	[9]	[9]	[9]	[9]	[9]	[9]	[9]

Part I　　　　　　　　　　　　**Writing**　　　　　　　　　　　**(30 minutes)**

Directions: *For this part, you are allowed thirty minutes to write a composition on the topic* **The Benefits of Volunteering**. *You should write no less than 120 words and you should base your composition on the outline (given in Chinese) below:*

1、阐明个人观点：做志愿者是有好处的；

2、论证上述观点；

3、总结做志愿者好处的观点的正确性。

The Benefits of Volunteering

_____

_____

_____

_____

_____

_____

_____

答题卡 1 （Answer Sheet 1）

_____

_____

_____

_____

_____

_____

_____

_____

_____

_____

Part II   Reading Comprehension (Skimming and Scanning)   (15 minutes)

1. [Y] [N] [NG]   2. [Y] [N] [NG]   3. [Y] [N] [NG]   4. [Y] [N] [NG]

5. [Y] [N] [NG]   6. [Y] [N] [NG]   7. [Y] [N] [NG]

必须使用黑色签字笔书写，在答题区域内作答，超出以下矩形边框限定区域的答案无效。

8. There are three components needed for ignition and combustion to occur: fuel, air and _____.

9. In a wildfire, the main variable in ignition time is the ration of the fuel's total surface area _____.

10. As far as a fire's behavior is concerned, the most unpredictable factor is _____.

答题卡 2（Answer Sheet 2）

Part III   Section A   Section B

11. [A] [B] [C] [D]   16. [A] [B] [C] [D]   21. [A] [B] [C] [D]   26. [A] [B] [C] [D]   31. [A] [B] [C] [D]

12. [A] [B] [C] [D]   17. [A] [B] [C] [D]   22. [A] [B] [C] [D]   27. [A] [B] [C] [D]   32. [A] [B] [C] [D]

13. [A] [B] [C] [D]   18. [A] [B] [C] [D]   23. [A] [B] [C] [D]   28. [A] [B] [C] [D]   33. [A] [B] [C] [D]

14. [A] [B] [C] [D]   19. [A] [B] [C] [D]   24. [A] [B] [C] [D]   29. [A] [B] [C] [D]   34. [A] [B] [C] [D]

15. [A] [B] [C] [D]   20. [A] [B] [C] [D]   25. [A] [B] [C] [D]   30. [A] [B] [C] [D]   35. [A] [B] [C] [D]

Part III   Section C

必须使用黑色签字笔书写，在答题区域内作答，超出以下矩形边框限定区域的答案无效。

Very high waves are (36)_____when they strike the land. Fortunately, this (37)_____ happens. One reason is that out at sea, waves moving in one direction almost always run into waves moving in a different direction. The two (38)_____of waves tend to cancel each other out. Another reason is that water is (39)_____near the shore. As a wave gets closer to land, the shallow (40)_____helps reduce its (41)_____.

But the power of waves striking the (42)_____can still be very great. During a winter gale, waves sometimes (43)_____ the shore with the force of 6,000 pounds for each square foot. That means a wave, 25 feet high and 500 feet along its face, may (44)_____. Yet (45)_____ During the most raging storms, (46)_____.

答题卡2（Answer sheet 2）

Part Ⅳ	Section A	Section B	Part Ⅴ	

Section A

47. [A][B][C][D][E][F][G][H][I][J][K][L][M][N][O]
48. [A][B][C][D][E][F][G][H][I][J][K][L][M][N][O]
49. [A][B][C][D][E][F][G][H][I][J][K][L][M][N][O]
50. [A][B][C][D][E][F][G][H][I][J][K][L][M][N][O]
51. [A][B][C][D][E][F][G][H][I][J][K][L][M][N][O]
52. [A][B][C][D][E][F][G][H][I][J][K][L][M][N][O]
53. [A][B][C][D][E][F][G][H][I][J][K][L][M][N][O]
54. [A][B][C][D][E][F][G][H][I][J][K][L][M][N][O]
55. [A][B][C][D][E][F][G][H][I][J][K][L][M][N][O]
56. [A][B][C][D][E][F][G][H][I][J][K][L][M][N][O]

Section B

57. [A][B][C][D]
58. [A][B][C][D]
59. [A][B][C][D]
60. [A][B][C][D]
61. [A][B][C][D]
62. [A][B][C][D]
63. [A][B][C][D]
64. [A][B][C][D]
65. [A][B][C][D]
66. [A][B][C][D]

Part Ⅴ

67. [A][B][C][D]   77. [A][B][C][D]
68. [A][B][C][D]   78.[A][B][C][D]
69. [A][B][C][D]   79. [A][B][C][D]
70. [A][B][C][D]   80. [A][B][C][D]
71. [A][B][C][D]   81. [A][B][C][D]
72. [A][B][C][D]   82. [A][B][C][D]
73. [A][B][C][D]   83. [A][B][C][D]
74. [A][B][C][D]   84. [A][B][C][D]
75. [A][B][C][D]   85. [A][B][C][D]
76. [A][B][C][D]   86. [A][B][C][D]

Part Ⅵ   Translation   (5 minutes)

必须使用黑色签字笔书写，在答题区域内作答，超出以下矩形边框限定区域的答案无效。

87. He continued speaking, _____ (不顾及)my feelings on the matter.

88. _____ (到目前为止)everything is all right.

89. Will the train arrive _____(准时)?

90. I can't go, she'll go _____me (由她代替我去).

91. She seems to do these things _____(有意地).

大学英语4级考试新题型模拟试卷 1—16

Model Test Two

Part I　　　　　　　　　　Writing　　　　　　　　(30 minutes)

注意：此部分试题在**答题卡 1** 上

Part Ⅱ　　Reading Comprehension (Skimming and Scanning)　　(15 minutes)

Directions: *In this part, you will have 15 minutes to go over the passage quickly and answer the questions on Answer Sheet 1.*

For questions 1~7, mark
　　Y (for YES)　　　　　　*if the statement agrees with the information given in the passage;*
　　N (for NO)　　　　　　*if the statement contradicts the information given in the passage;*
　　NG (for NOT GIVEN)　*if the information is not given in the passage.*

For questions 8~10, complete the sentence with the information given in the passage.

注意：此部分试题请在**答题卡 1** 上作答

SAVING LANGUAGE

There is nothing unusual about a single language dying. Communities have come and gone throughout history and with them their language. But what is happening today is extraordinary, judged by the standards of the past. It is language extinction on a large scale. According to the best estimate, there are some 6,000 languages in the world. Of course, about half are going to die out in the course of the next century: that's 3,000 languages in 1,200 months. On average, there is a language dying out somewhere in the world every two weeks or so.

How do we know? In the course of the past two or three decades, linguists all over the world have been gathering comparative data. If they find a language with just a few speakers left, and nobody is bothering to pass the language on to the children, they conclude that language is bound to die out soon. And we have to draw the same conclusion if a language has less than 100 speakers. It is not likely to last very long. A 1999 survey shows that 97 percent of the world's languages are spoken by just four percent of the people.

It is too late to do anything to help many languages, where the speakers are too few or too old, and where the community is too busy just trying to survive to care about their language. But many languages are not in such a serious position. Often, where languages are seriously endangered, there are things that can be done to give new life to them. It is called revitalization.

Once a community realizes that its language is in danger, it can start to introduce measures which can genuinely revitalize. The community itself must want to save its language. The culture of which it is a part must need to have a respect for minority languages. There need to be funding to support courses, materials, and teachers. And there need to be linguists to get on with the basic task of putting the language down on paper. That's the bottom line: getting the language documented—recorded, analyzed, written down. People must be able to read and write down. People must be able to read and write if they and their language are to have a future in an increasingly computer-literate civilization.

But can we save a few thousand languages, just like that? Yes, if the will and funding were available. It is not cheap getting linguists into the field, training local analysts, supporting the community with language resources and teachers, compiling grammars and dictionaries, writing materials for use in schools. It takes time, lots of it, to revitalize an endangered language. Conditions vary so much that it is difficult to generalize, but a figure of $900 million.

There are some famous cases which illustrate what can be done. Welsh, alone among the Celtic languages, is not only stooping its steady decline towards extinction but showing signs of real growth. Two language acts protect the status of Welsh now, and its presence is increasingly in evidence wherever you travel in Wales.

On the other side of the world, Maori in New Zealand has been maintained by a system of so-called "language nests", first introduced in 1982. These are organizations which provide children under five with a domestic setting in which they are all intensively exposed to the language. The staffs are all Maori speakers from the local community. The hope is that the children will keep their Maori skills alive after leaving the nests, and that as they grow older they will in turn become role models to a new generation of young children. These are cases like this all over the world. And when the reviving language is associated with a degree of political autonomy, the growth can be especially striking, as shown by Faroese, spoken in the Faeroe Island, after the islanders received a measure of autonomy from Denmark.

In Switzerland, Romansch was facing a difficult situation, spoken in five very different dialects, with small and diminishing numbers, as young people left their community numbers in the German-speaking cities. The solution here was the creation in the 1980s of a unified written language for all these dialects. Romansch Grischun, as it is now called, has official status in parts of Switzerland, and is being increasingly used in spoken form on radio and television.

A language can be brought back from the very brink of extinction. The Ainu language of Japan, after many years of neglect and repression, had reached a stage where there were only eight fluent speakers left, all elderly. However, new government policies brought fresh attitudes and a positive interest in survival. Several "semi-speakers"—people who become unwilling to speak Ainu because of the negative attitudes by Japanese speakers—were prompted to become active speakers again. This is fresh interest now and the language is more publicly available than it has been for years.

If good descriptions and materials are available, even extinct languages can be revived. Kaurna, from South Australia, is an example. This language had been extinct for about a century, but had been quite well documented. So, when a strong movement grew for its revival, it was possible to reconstruct it. The revised language is not the same as the original, of course. It lacks the range that the original had, and much of the old vocabulary. But it can nonetheless act as a badge of present-day identity for its people. And as long as people continue to value it as a true marker of their identity, and are prepared to keep using it, it will develop new functions and new vocabulary, as many other living language would do.

It is too soon to predict the future of these revived languages, but in some parts of the world they are attracting precisely the range of positive attitudes and grass roots support which are the preconditions for language survival. In such unexpected but heart-warming ways might we see the grand total of languages in the world increased.

Questions:

1. The rate at which languages are becoming extinct has increased.

2. Research on the subject of language extinction began in the 1990s.

3. In order to survive, a language needs to be spoken by more than 100 people.

4. Language extinct more quickly in certain parts of the world than in others.

5. The small community whose language is under threat can take measures to revitalize the language.

6. A few thousand languages can be saved if enough funds are raised to do so.

7. An extinct language can never be revived no matter what you do about it.

8~10题请见**答题卡 1**

Part III   **Listening Comprehension**   (35 minutes)

Section A

Directions: *In this section, you will hear 8 short and 2 long conversations. At the end of each conversation , one or more questions will be asked about what was said. Both the conversation and the question will be spoken only once. After each question there will be a pause. During the pause, you must read the four choices marked A), B), C) and D), and decide which is the best answer. Then mark the corresponding letter on the **Answer Sheet** 2 with a single line through the center.*

注意：此部分试题请在**答题卡 2** 上作答

11. A) The man doesn't want to see Mr. Williams.

    B) Mr. Jones is in an inferior position than Mr. Williams.

    C) Mr. Jones used to be in charge.

    D) Mr. Williams doesn't want to do tomorrow.

12. A) They need to make more efforts.

    B) They'll have more work to do tomorrow.

    C) The others have done the greater part of it.

    D) They've finished more than half of it.

13. A) She was feeling very sorry.　　　B) She felt a bit annoyed.

    C) She was in a hurry.　　　D) She was in her office.

14. A) Jane was telling a lie.

    B) The woman wasn't being sincere.

    C) Jane has already come back from Paris.

    D) Jane wasn't in Paris that day.

15. A) The knife belongs to him.

    B) Bob should mind his own business.

    C) The man once borrowed Bob's knife.

    D) Bob's knife isn't as good as that of the man.

16. A) He'll miss the meeting that afternoon.　　B) He'll have an appointment with the host.

    C) He won't miss the meeting.　　D) He is very hardworking.

17. A) Because she won't fulfill her promise.

    B) Because her mother would be very angry.

    C) Because she can't finish the job ahead of schedule.

    D) Because she would be the last to finish the job.

18. A) He always talks on the phone for that long if it's toll free.

    B) They have so much free time to talk on the phone for that long.

    C) They talked on the phone for too long.

    D) He wants to know what they talked about.

Questions 19 to 22 are based on the conversation you have just heard.

19. A) Buying a pair of Adidas tennis shoes.

    B) Asking her father about Adidas shoes.

    C) Discussing with her father about Adidas shoes.

    D) Joining the tennis club in school.

20. A) They don't help sports players at all.

    B) They don't live up to their fame.

    C) They may be comfortable but are too expensive.

    D) They are good for track and field sports but not for ball games.

21. A) He and his friends have never worn Adidas.

    B) Adidas is just for great players like the Chicago Bulls.

    C) Adidas helps sports players do better.

    D) He has always wanted to wear Adidas shoes.

22. A) He thinks Adidas would cost him quite a lot of money.

   B) He doesn't think Joyce will run.

   C) He doesn't think the shoes will help Joyce in the games.

   D) He doesn't think Joyce will play sports for long.

Questions 23 to 25 are based on the conversation you have just heard.

23. A) Tourist and guide.   B) Teacher and student.

   C) Travel agent and tourist.   D) Customer and shop assistant.

24. A) He wants to find a guide to show him around the city.

   B) He will travel to another city in a couple of days.

   C) He is buying some books with a lot of pictures and a map.

   D) He is quite new here.

25. A) A specialized one with a map.

   B) A specialized one with a lot of pictures.

   C) A general one with pictures.

   D) An easy one with pictures and a map.

Section B

Directions: *In this section, you will hear 3 short passages. At the end of each passage, you will hear some questions. Both the passage and the questions will be spoken only once. After you hear a question, you must choose the best answer from the four choices marked A), B), C) and D). Then mark the corresponding letter on the **Answer Sheet** 2 with a single line through the center.*

注意：此部分试题请在**答题卡 2**上作答

Passage One

Questions 26 to 28 are based on the passage you have just heard.

26. A) He was a tax collector.

   B) He was a government official.

   C) He was once a friend of the ruler.

   D) He was once a school teacher in India.

27. A) To reward outstanding tax collectors.

   B) To declare new ways of collecting tax.

   C) To collect money from the persons invited.

   D) To entertain those who had made great contributions to the government.

28. A) They were excused from paying income tax.

   B) They were given some silver and gold coins by the ruler.

   C) They tried to collect more money than the ruler asked for.

   D) They enjoyed being invited to dinner at the ruler's palace.

Passage Two

Questions 29 to 31 are based on the passage you have just heard.

29. A) They liked traveling.

    B) The reasons are unknown.

    C) They were driven out of their homes.

    D) They wanted to find a better place to live in.

30. A) They are unfriendly to Gypsies.

    B) They admire the musical talent of the Gypsies.

    C) They are envious of Gypsies.

    D) They try to put up with Gypsies.

31. A) They are now taught in their own language.

    B) They are now allowed to attend local schools.

    C) Special schools have been set up for them.

    D) Permanent homes have been built for them.

Passage Three

Questions 32 to 35 are based on the passage you have just heard.

32. A) The causes are familiar. B) The causes are not well understood.

    C) The causes are obvious. D) The causes are very complicated.

33. A) Improved highway design. B) Better public transportation.

    C) Regular driver training. D) Stricter traffic regulations.

34. A) Highway crime. B) Drivers' errors.

    C) Poor traffic control. D) Confusing road signs.

35. A) Increasing people's awareness of traffic problems.

    B) Enhancing drivers' sense of responsibility.

    C) Building more highways.

    D) Designing better cars.

Section C

Directions: *In this section, you will hear a passage three times. When the passage is read for the first time, you should listen carefully for its general idea. When the passage is read for the second time, you are required to fill in the blanks numbered from 36 to 43 with the exact words you have just heard. For blanks numbered from 44 to 46 you are required to fill in the missing information. You can either use the exact words you have just heard or write down the main points in your own words. Finally, when the passage is read for the third time, you should check what you have written.*

注意：此部分试题在**答题卡 2** 上，请在**答题卡 2** 上作答

Part IV Reading Comprehension (Reading in Depth) (25 minutes)

Section A

Directions: *In this section, there is a passage with ten blanks. You are required to select one word for each blank from a list of choices given in a word bank following the passage. Read the passage through carefully before making your choices. Each choice in the bank is identified by a letter. Please mark the corresponding letter for each item on **Answer Sheet 2** with a single line through the centre. **You may not use any of the words in the bank more than once.***

Questions 47 to 56 are based on the following passage.

As war spreads to many corners of the globe, children sadly have been drawn into the center of conflicts. In Afghanistan, Bosnia, and Colombia, however, groups of children have been taking part in peace education   47   . The children, after learning to resolve conflicts, took on the   48   of peacemakers. The Children's Movement for Peace in Colombia was even nominated (提名) for the Nobel Peace Prize in 1998. Groups of children   49   as peacemakers studied human rights and poverty issues in Colombia, eventually forming a group with five other schools in Bogota known as The Schools of Peace.

The classroom   50   opportunities for children to replace angry, violent behaviors with   51   , peaceful ones. It is in the classroom that caring and respect for each person empowers children to take a step   52   toward becoming peacemakers. Fortunately, educators have access to many online resources that are   53   useful when helping children along the path to peace. The Young Peacemakers Club, started in 1992, provides a Website with resources for teachers and   54   on starting a Kindness Campaign. The World Centers of Compassion for Children International call attention to children's rights and how to help the   55   of war. Starting a Peacemakers' Club is a praiseworthy venture for a class and one that could spread to other classrooms and ideally affect the culture of the   56   school.

注意：此部分试题请在**答题卡 2**上作答

A) acting	I) information
B) assuming	J) offers
C) comprehensive	K) projects
D) cooperative	L) respectively
E) entire	M) role
F) especially	N) technology
G) forward	O) victims
H) images	

Section B

Directions: *There are 2 passages in this section. Each passage is followed by some questions or unfinished statements. For each of them there are four choices marked A), B), C) and D). You should decide on the best choice and mark the corresponding letter on the **Answer Sheet 2** with a single line through the centre.*

Passage one

Questions 57 to 61 are based on the following passage.

Drunken driving, sometimes called America's socially accepted form of murder, has become a national *epidemic* (流行病). Every hour of every day about three Americans on average are killed by drunken drivers, adding up to an incredible 350,000 over the past decade.

A drunken driver is usually defined as one with a 0.10 blood alcohol content or roughly three beers, glasses of wine or shots of whisky drunk within two hours. Heavy drinking used to be an acceptable part of the American alcohol image and judges were *lenient* (宽容的) in most courts, but the drunken slaughter has recently caused so many well-publicized tragedies, especially involving young children, that public opinion is no longer so tolerant.

Twenty states have raised the legal drinking age to 21, reversing a trend in the 1960's to reduce it to 18. After New Jersey lowered it to 18, the number of people killed by 18-to-20-year-old drivers more than doubled, so the state recently upped it back to 21.

Reformers, however, fear raising the drinking age will have little effect unless accompanied by educational programs to help young people to develop "responsible attitudes" about drinking and teach them to resist peer pressure to drink.

Tough new laws have led to increased arrests and tests and in many areas already, to a marked decline in fatalities. Some states are also penalizing bars for serving customers too many drinks.

As the fatalities continue to occur daily in every state, some Americans are even beginning to speak well of the 13 years' national prohibition of alcohol that began in 1919, which President Hoover called the "noble experiment". They forget that legal prohibition didn't stop drinking, but encouraged political corruption and organized crime. As with the booming drug trade generally, there is no easy solution.

注意：此部分试题请在答题卡 2 上作答

57. Which of the following best concludes the main idea of the passage?

A) Drunken driving has caused numerous fatalities in the United States.

B) It's recommendable to prohibit alcohol drinking around the United States.

C) The American society is trying hard to prevent drunken driving.

D) Drunken driving has become a national epidemic in the United States.

58. Which of the following four drivers can be defined as an illegal driver?

    A) A sixteen-year-old boy who drank a glass of wine three hours ago.

    B) An old lady who took four shots of whisky in yesterday's party.

    C) A policeman who likes alcohol very much.

    D) A pregnant woman who drank a beer an hour ago.

59. In reformers' opinion, _____ is the most effective way to stop youngsters from drinking alcohol.

    A) raising the legal drinking age from 18 to 21

    B) forcing teenagers to obey disciplines

    C) developing young people's sense of responsibility

    D) pressing teenagers to take soft drinks

60. The rule that only people above 21 years of age can drink _____.

    A) is a new law promoted by the twenty states

    B) had been once adopted before the 1960's

    C) has been enforced since the prohibition of alcohol

    D) will be carried out all over the country

61. What is the author's attitude toward all the laws against drunken driving?

    A) Optimistic.   B) Pessimistic.

    C) Indifferent.   D) Ironic.

Passage Two

Questions 62 to 66 are based on the following passage.

The economic effects are easy to see. Since 1978, some 43 million jobs have been lost largely to forms of technology—either to robotics directly or to computers that are doing what they are supposed to be doing, being labor-saving devices. Today, there is no such thing as a lifetime job; there is no such thing as a career for most people anymore. The jobs that are not done away with are being deskilled, or they are disposable jobs. Even for those jobs that many of you may feel secure with, there are people who are working on what are called "expert systems" to be able to take jobs away from doctors and judges and lawyers. The machine is capable of *shredding* (切碎) these jobs as well.

But it's not just the jobs. The economy of jobs and services is trivial compared to the "Nintendo capitalism" that now operates in the world. Four trillion dollars a day is *shuffled* (流通) around the earth. The inevitable result of a Nintendo economy—pulling itself apart, losing jobs, insecure—is the *shriveling* (萎缩) of the society in which it exists. What we have is an *apartheid*(种族隔离的)society, with growing gaps between the rich and poor, and the rich spending a lot of time protecting themselves from the effects of the poor.

A further result of information technology—something that nobody seems to wish to pay much attention to is the shredding everywhere of the natural world. Forget about the number of toxins that go into producing these computers, and the resources that go into producing them, such that 40,000 pounds of resources are necessary for a four-pound laptop. That's trivial compared with the direct effect that computers and the industrial system as a result have on the atmosphere and climate, the pollution of air and water.

The development in technology does not always bring human beings goods; there is bad news too. But most people are ignorant of the drawback of the new technology at first. In this century, however, the development in science and technology really aroused people's attention of the weak points. But the technology has an even darker effect, because it is enabling us to conquer nature. Industrial society is waging a war of the techno-sphere against the biosphere. That is the Third World War. The bad news is that we are winning that war.

注意：此部分试题请在**答题卡 2** 上作答

62. According to the passage, information technology brings hazard to_____.

    A) human society and natural environment.

    B) natural environment and economy

    C) domestic economy and human society

    D) economy, human society and environment

63. From the context, we can infer that "Nintendo capitalism" means_____.

    A) a capitalism that is prosperous

    B) a capitalism that leads to destruction

    C) a worship of capitalism

    D) a worship of technology

64. The statement made by the author that what we have is an apartheid society because_____.

    A) the rich are richer and the poor are poorer

    B) the white are prejudiced against the black

    C) a lot of jobs have been lost to high technology

    D) the society is gradually shriveling

65. According to the third paragraph, what has caused the pollution of air and water?

    A) The resources used to produce computers.

    B) The four-pound laptop.

    C) Computers and industrial system.

    D) Shredding of the natural world.

66. When it comes to the term "the Third World War", the author tries to imply_____.

    A) human conquering of nature.

    B) information technology's destruction of natural environment.

    C) technology's control over nature.

    D) technology's conquering of human society and nature.

Part V   Error Correction   (15 minutes)

Directions: *The passage contains TEN errors. Each indicated line contains a maximum of ONE error. In each case, only ONE word involved: You should proofread the passage and correct it in the following way.*

For a wrong word, underline the wrong word and write the correct one in the blank provided at the end of the line.

For a missing word, mark the position of the missing word with a " ^ " sign and write the word you believe to be missing in the blank provided at the end of the line.

For an unnecessary word, cross the unnecessary word with a slash " / " and put the word in the blank provided at the end of the line.

EXAMPLE

When ^art museum wants a new exhibit,	(1) an
it never buys things in finished form and hangs	(2) never
them on the wall. When a natural history museum	
wants an exhibition, it must often build it.	(3) exhibit

注意：此部分试题在**答题卡 2** 上，请在**答题卡 2** 上作答

Part VI   Translation   (5 minutes)

Directions: *Complete the sentences on **Answer Sheet 2** by translating the Chinese given in brackets into English.*

注意：此部分试题在**答题卡 2** 上，请在**答题卡 2** 上作答

答题卡 1 （Answer Sheet 1）

Part I   **Writing**   **(30 minutes)**

Directions: *For this part, you are allowed thirty minutes to write a composition on the topic* **The Importance of Reading Classics**. *You should write no less than 120 words and you should base your composition on the outline (given in Chinese) below:*

1. 阐明个人观点：阅读名著的重要性；
2. 论证上述观点；
3. 总结阅读名著重要性这一观点的正确性。

The Importance of Reading Classics

答题卡 1 （Answer Sheet 1）

_____

_____

_____

_____

_____

_____

_____

_____

_____

_____

Part Ⅱ   **Reading Comprehension (Skimming and Scanning)**   **(15 minutes)**

1. [Y] [N] [NG]   2. [Y] [N] [NG]   3. [Y] [N] [NG]   4. [Y] [N] [NG]

5. [Y] [N] [NG]   6. [Y] [N] [NG]   7. [Y] [N] [NG]

必须使用黑色签字笔书写，在答题区域内作答，超出以下矩形边框限定区域的答案无效。

8. Romansch Grischun is a (an) _____ language in parts of Switzerland.

9. The example of Ainu illustrates that a language can be saved _____.

10. The preconditions for a language to survive is the people's_____.

答题卡 2（Answer Sheet 2）

准	考	证	号											
[0]	[0]	[0]	[0]	[0]	[0]	[0]	[0]	[0]	[0]	[0]	[0]	[0]	[0]	[0]
[1]	[1]	[1]	[1]	[1]	[1]	[1]	[1]	[1]	[1]	[1]	[1]	[1]	[1]	[1]
[2]	[2]	[2]	[2]	[2]	[2]	[2]	[2]	[2]	[2]	[2]	[2]	[2]	[2]	[2]
[3]	[3]	[3]	[3]	[3]	[3]	[3]	[3]	[3]	[3]	[3]	[3]	[3]	[3]	[3]
[4]	[4]	[4]	[4]	[4]	[4]	[4]	[4]	[4]	[4]	[4]	[4]	[4]	[4]	[4]
[5]	[5]	[5]	[5]	[5]	[5]	[5]	[5]	[5]	[5]	[5]	[5]	[5]	[5]	[5]
[6]	[6]	[6]	[6]	[6]	[6]	[6]	[6]	[6]	[6]	[6]	[6]	[6]	[6]	[6]
[7]	[7]	[7]	[7]	[7]	[7]	[7]	[7]	[7]	[7]	[7]	[7]	[7]	[7]	[7]
[8]	[8]	[8]	[8]	[8]	[8]	[8]	[8]	[8]	[8]	[8]	[8]	[8]	[8]	[8]
[9]	[9]	[9]	[9]	[9]	[9]	[9]	[9]	[9]	[9]	[9]	[9]	[9]	[9]	[9]

Part III   Section A   Section B

11. [A] [B] [C] [D]   16. [A] [B] [C] [D]   21. [A] [B] [C] [D]   26. [A] [B] [C] [D]   31. [A] [B] [C] [D]
12. [A] [B] [C] [D]   17. [A] [B] [C] [D]   22. [A] [B] [C] [D]   27. [A] [B] [C] [D]   32. [A] [B] [C] [D]
13. [A] [B] [C] [D]   18. [A] [B] [C] [D]   23. [A] [B] [C] [D]   28. [A] [B] [C] [D]   33. [A] [B] [C] [D]
14. [A] [B] [C] [D]   19. [A] [B] [C] [D]   24. [A] [B] [C] [D]   29. [A] [B] [C] [D]   34. [A] [B] [C] [D]
15. [A] [B] [C] [D]   20. [A] [B] [C] [D]   25. [A] [B] [C] [D]   30. [A] [B] [C] [D]   35. [A] [B] [C] [D]

Part III   Section C

必须使用黑色签字笔书写，在答题区域内作答，超出以下矩形边框限定区域的答案无效。

Russia is the largest economic power that is not a member of the World Trade Organization. But that may change. Last Friday, the European Union said it would support Russia's (36) _____ to become a WTO member.

Representatives of the European Union met with Russian (37) _____ in Moscow. They signed a trade agreement that took six years to (38) _____.

Russia called the trade agreement (39) _____. It agreed to slowly increase fuel prices within the country. It also agreed to permit (40) _____ in its communications industry and to remove some barriers to trade.

In (41) _____ for European support to join the WTO, Russian President Putin said that Russia would speed up the (42) _____ to approve the Kyoto Protocol, and international (43) _____ agreement to reduce the production of harmful industrial gases. (44) _____.

Russia had signed the Kyoto Protocol, but has not yet approved it. The agreement takes effect when it has been approved by nations that produce at least 55 percent of the world's greenhouse gases. (45) _____. The United States, the world's biggest producer, withdrew from the Kyoto Protocol after President Bush took office in 2001. So Russia's approval is required to put the Kyoto Protocol into effect. (46) _____, Russia must still reach agreements with China, Japan, South Korea besides the United States.

答题卡 2（Answer sheet 2）

Part Ⅳ　　　Section A　　　　　　　　　　Section B

47. [A][B][C][D][E][F][G][H][I][J][K][L][M][N][O]　　57. [A][B][C][D]
48. [A][B][C][D][E][F][G][H][I][J][K][L][M][N][O]　　58. [A][B][C][D]
49. [A][B][C][D][E][F][G][H][I][J][K][L][M][N][O]　　59. [A][B][C][D]
50. [A][B][C][D][E][F][G][H][I][J][K][L][M][N][O]　　60. [A][B][C][D]
51. [A][B][C][D][E][F][G][H][I][J][K][L][M][N][O]　　61. [A][B][C][D]
52. [A][B][C][D][E][F][G][H][I][J][K][L][M][N][O]　　62. [A][B][C][D]
53. [A][B][C][D][E][F][G][H][I][J][K][L][M][N][O]　　63. [A][B][C][D]
54. [A][B][C][D][E][F][G][H][I][J][K][L][M][N][O]　　64. [A][B][C][D]
55. [A][B][C][D][E][F][G][H][I][J][K][L][M][N][O]　　65. [A][B][C][D]
56. [A][B][C][D][E][F][G][H][I][J][K][L][M][N][O]　　66. [A][B][C][D]

Part V　　　　　　　　**Error Correction**　　　　　　　　**(15 minutes)**

　　The first man known to use a signal other than a bonfire used a
chandelier. He was lord of a castle that stood near a rocky seacoast. He hangs　67_____
the chandelier, containing many large tallow candles, in the highest tower of
his castle. Thus he warned passing ship from the danger along the coast.　68_____

　　Candles soon became the common fuel for signal lights. They were later
replaced by oil lamps, that could burn larger and brighter. Kerosene and gas　69_____
lamps also tried. These are still in use now in some smaller lighthouses. But　70_____
today most lighthouses sent electric light blazing out over the sea.　71_____

　　The ancient fire signals only say "Danger! Keep off!" But the modern
lighthouse also identifies it in a code known to all shipping. Most of the great　72_____
lights have their own special signals. The light may be one that blinks—as a　73_____
giant firefly in the night. Or it may be a revolved light that is red then green.　74_____
Or it may be only white. But however the signal, it is sent very regularly. A　75_____
ship within its ranges is never at a loss to know which lighthouse it is, and
where it is being located.　76_____

Part Ⅵ   Translation   (5 minutes)

必须使用黑色签字笔书写，在答题区域内作答，超出以下矩形边框限定区域的答案无效。

77. _____ (多亏你们的帮助), we accomplished the task in time.

78. _____ (这到底是谁告诉你的？) That is not true.

79. _____ (假如遇到)fire, ring the alarm bell.

80. It came to my ears _____ (偶然).

81. I saw Tom and _____ (他一次拿六个盒子).

Model Test Three

Part I	Writing	(30 minutes)

注意：此部分试题在**答题卡 1** 上

Part II	Reading Comprehension (Skimming and Scanning)	(15 minutes)

Directions: *In this part, you will have 15 minutes to go over the passage quickly and answer the questions on **Answer Sheet 1**.*

For questions 1~7, mark

Y *(for YES)*   *if the statement agrees with the information given in the passage;*

N *(for NO)*   *if the statement contradicts the information given in the passage;*

NG *(for NOT GIVEN)*   *if the information is not given in the passage.*

For questions 8~10, complete the sentence with the information given in the passage.

New York

As a travel destination, New York has something to offer almost every visitor. Though tourism has dropped since September 11, 2001, there are still lots of reasons to visit what many consider the greatest city in the world.

City Overview

New York City (NYC) is located on the eastern Atlantic coast of the United States. It rests at the mouth of the Hudson River. The city is often referred to as a "city of island". Greater NYC is made up of five distinct areas called boroughs. These boroughs include Manhattan, Brooklyn, Queens Island and the Bronx. The boroughs are separated from each other by various bodies of water and are connected by subways, bridges and tunnels.

When people refer to New York City, they are usually talking about Manhattan. Most of NYC's main attractions are located in this borough and the majority of visitors spend most of their vacation here.

A Short History of the Big Apple

No discussion about New York City would be complete without asking why New York is referred to as "the Big Apple". Like many things about New York, you'll probably get a different answer depending on who you ask. According to the Museum of the City of New York, it is believed that in the 1920s, a sportswriter overheard stable hands in New Orleans refer to New York City's racetracks as "the Big

Apple". The phrase was most widely used by jazz musicians during the 1930s and 1940s. They adopted the term to refer to New York City, and especially Harlem, as the jazz capital of the world.

The Italian navigator Giovanni da Verrazano may have been the first European to explore the New York region in 1524. More than 80 years later, Englishman Henry Hudson sailed up the river that now bears his name. But it was Dutch settlements that truly started the city. In 1624, the town of New Amsterdam was established on lower Manhattan. Two years later, according to local legend, Dutchman Peter Minuit purchased the island of Manhattan from the local Native Americans for 60 guilders (about $24) worth of goods.

Few people realize that New York was briefly the U.S. capital from 1789 to 1790 and was the capital of New York State until 1797. By 1790, it was the largest U.S. city. In 1825, the opening of the Erie Canal, which linked New York with the Great Lakes, led to continued expansion.

A charter was adopted in 1898 incorporating all five boroughs into Great New York. New York has always been and remains a city of immigrants. Patterns of immigration are integral to the city's history and landscape. Immigration, mainly from Europe, swelled the city's population in the late 19th and early 20th centuries. After World War II, many African-Americans from South, Puerto Ricans, and Latin Americans migrated to the city as well. Because of the variety of immigrant group, both historically and currently, New York is often referred as a true "melting pot".

Getting Around: By Foot

The absolute best way to get around New York, and the one you will probably be using most, is walking. Remember, the city is only 13 miles long. On a day with good weather, walking is a great option. The excitement of New York on foot is that you never know what interesting thing you will see as you head from one destination to another. If you're on a schedule, keep in mind that distances are not as close as they might seem and take into account the extra time it takes to stop at every street crosswalk. Getting from the easternmost side of Manhattan to the westernmost side can take quite a while.

The Subway

If you're looking to save some time, this is where one of the three excellent New York public transportation systems comes in handy. They are all run by the city's Metropolitan Transportation Authority. According to NYC & Company, the city's official visitor's bureau, the 714—mile New York City subway system has 468 stations serving 24 routes—more than any other system in the world. It operates 24 hours a day, is safe, and is used daily by more than 3.5 million people.

The main thing to remember when using the subway system is to make sure you get on the correct train. Uptown trains head north, downtown and Brooklyn borough trains head south. Express trains make every stop. New Yorkers and tourists alike have hopped on the wrong train and ended up in an unknown area. If this happens, simply hop the next train back the way you arrived. To avoid these problems, when in doubt, always take a local train.

Buses & Taxis

The final two systems of public transportation are buses and taxis. Buses tend to be very slow because of New York traffic, but they can give you great views of the city streets. Buses run north and south as well as east and west. Just like the subway, the bus system has its own map and routes. Most free subway maps also include a bus map.

Taxis are usually quicker to navigate the city streets than buses. But be prepared to pay for that convenience. Taxis are expensive. A trip from the Upper West Side, for example, to the lower East Side can cost upwards of $12, not including the driver's tip. There is an automatic $2 charge on all cab rides and all taxi drivers expect some sort of gratuity.

Main Attractions

Here is a small sampling of some of the main attractions NYC is best known for.

Statue of Liberty—Few New York sites are as awe—inspiring as this one. A century ago, Lady Liberty held up her torch to welcome immigrants to America. Today, you can climb 354 steps to look out from her crown to see both the New York and New Jersey coasts.

Times Square—New Yorkers call this intersection of Broadway and 42nd street the "Crossroad of the World". It is the most recognized intersection on earth—millions of people see it on television every New Year's Eve. Some people say it's the best place in New York to people—watch. At night, the illuminated signs in Times Square make an amazing light show.

Empire States Building—built in 1931, this skyscraper was the tallest in the world for half a century. You'll get a great view of the city from the art deco tower's observation deck.

Central Park—Who would have thought that a city filled with people, traffic and skyscrapers, could offer visitors such an incredible natural oasis? The park is full of rolling meadows, trees, water bodies and stone bridge. The best part ? It's all free.

Metropolitan Museum of Art—If you see only one museum in New York City, the Met, as it is known, should be the one. The museum houses over two million works of art ranging from Egyptian to Medieval to 20th Century.

United Nations—You can't miss the 188 nations' flags high above First Avenue in front of the headquarters of this international organization. Tours take you through the Security Council and General Assembly Halls.

注意：此部分试题请在答题卡 1 上作答

Questions

1. People usually use the name New York City for Brooklyn.

2. There is a unanimous agreement as to why New York is referred to as "the Big Apple".

3. New York was the U.S. capital from 1789 to 1790.

4. New York City subway system ranks first in the world in terms of the number of stations.

5. Times Square is the most recognized intersection on earth.

6. If you see only museum in New York City, you should visit the American Museum of Natural History.

7. This passage is mainly about getting around New York City.

8～10 题请见**答题卡 1**

Part Ⅲ   Listening Comprehension   (35 minutes)

Section A

Directions: *In this section, you will hear 8 short and 2 long conversations. At the end of each conversation , one or more questions will be asked about what was said. Both the conversation and the question will be spoken only once. After each question there will be a pause. During the pause, you must read the four choices marked A), B), C) and D), and decide which is the best answer. Then mark the corresponding letter on the **Answer Sheet 2** with a single line through the center.*

注意：此部分试题请在**答题卡 2** 上作答

11. A) At the department store.   B) At the office.
    C) In the restaurant.   D) In the drug store.

12. A) He is upset.   B) He is disappointed.
    C) He is confident.   D) He is worried.

13. A) It's too high.   B) It's acceptable.
    C) It's cheap indeed.   D) The woman should have bargained for it.

14. A) At two o'clock.   B) At three o'clock.
    C) At four o'clock.   D) At five o'clock.

15. A) Shop assistant and customer.   B) Post clerk and customer.
    C) Store keeper and customer.   D) Waitress and customer.

16. A) To park the car.   B) To take pictures.
    C) To take her to the park.   D) To have a look at the sight.

17. A) His girlfriend complained of his going to the party without her.

    B) He was together with her girlfriend yesterday.

    C) He has been busy dating his girlfriend these days.

    D) He brought his girlfriend to the party.

18. A) She regretted having bought the second-hand car.

    B) It is unnecessary to rent another house.

    C) They should sell their second-hand car and buy a new one.

    D) They can afford a second-hand car.

Questions 19 to 22 are based on the conversation you have just heard.

19. A) At a television studio.   B) On a radio program.

    C) In a job interview.   D) In a factory.

20. A) He thinks people are more likely to buy music than books.

    B) He believes that there is a demand for books on music.

    C) He thinks that there isn't much future for e-books.

    D) He thinks that there is a good chance that the business will develop.

21. A) She seems rather sympathetic.   B) She appears rude.

    C) She seems to be hostile.   D) She appears to find them very interesting.

22. A) The prize was worth $ 100,000.

    B) It was for the best book published in electronic form.

    C) It was for the best book at the Frankfurt Book Fair.

    D) It was for the best software company.

Questions 23 to 25 are based on the conversation you have just heard.

23. A) He was fired by the bandleader.   B) He didn't have enough time.

    C) He had an offer to join a better band.   D) He felt he wasn't talented enough.

24. A) He played the trumpet.   B) He led the band.

    C) He was the drummer.   D) He played the saxophone.

25. A) Visit him at home.   B) Telephone him.

    C) See him perform.   D) Wait for a call from him.

Section B

Directions: *In this section, you will hear 3 short passages. At the end of each passage, you will hear some questions. Both the passage and the questions will be spoken only once. After you hear a question, you must choose the best answer from the four choices marked A), B), C) and D). Then mark the corresponding letter on the **Answer Sheet 2** with a single line through the center.*

Passage One

Questions 26 to 29 are based on the passage you have just heard.

26. A) A car outside the supermarket.   B) A car at the bottom of the hill.
    C) Paul's car.   D) The sports car.

27. A) Inside the car.   B) At the foot of the hill.
    C) In the garage.   D) In the supermarket.

28. A) The driver of the sports car.   B) The two girls inside the car.
    C) The man standing nearby.   D) The salesman from London.

29. A) Nobody.   B) The two girls.
    C) The bus driver.   D) Paul.

Passage Two

Questions 30 to 32 are based on the passage you have just heard.

30. A) His friend gave him the wrong key.

    B) He didn't know where the back door was.

    C) He couldn't find the key to his mailbox.

    D) It was too dark to put the key in the lock.

31. A) It was getting dark.

    B) He was afraid of being blamed by his friend.

    C) The birds might have flown away.

    D) His friend would arrive any time.

32. A) He looked silly with only one leg inside the window.

    B) He knew the policeman wouldn't believe him.

    C) The torch light made him look very foolish.

    D) He realized that he had made a mistake.

Passage Three

Questions 33 to 35 are based on the passage you have just heard.

33. A) The threat of poisonous desert animals and plants.

    B) The exhaustion of energy resources.

    C) The destruction of oil wells.

    D) The spread of the black powder from the fires.

34. A) The underground oil resources have not been affected.

    B) Most of the desert animals and plants have managed to survive.

    C) The oil lakes soon dried up and stopped evaporating.

    D) The underground water resources have not been polluted.

35. A) To restore the normal production of the oil wells.

 B) To estimate the losses caused by the fires.

 C) To remove the oil left in the desert.

 D) To use the oil left in the oil lakes.

Section C

Directions: *In this section, you will hear a passage three times. When the passage is read for the first time, you should listen carefully for its general idea. When the passage is read for the second time, you are required to fill in the blanks numbered from 36 to 43 with the exact words you have just heard. For blanks numbered from 44 to 46 you are required to fill in the missing information. You can either use the exact words you have just heard or write down the main points in your own words. Finally, when the passage is read for the third time, you should check what you have written.*

注意: 此部分试题在**答题卡 2 上**，请在**答题卡 2 上作答**

Part IV   Reading Comprehension (Reading in Depth)   (25 minutes)

Section A

Directions: *In this section, there is a passage with ten blanks. You are required to select one word for each blank from a list of choices given in a word bank following the passage. Read the passage through carefully before making your choices. Each choice in the bank is identified by a letter. Please mark the corresponding letter for each item on **Answer Sheet 2** with a single line through the centre. **You may not use any of the words in the bank more than once.***

Questions 47 to 56 are based on the following passage.

Asked to name their favorite city, many Americans would select San Francisco. San Francisco began as a __47__ Spanish outpost located on a magnificent bay. The town was little more than a village serving ranchers when the United States took __48__ of it in 1846 during the war with Mexico.

San Francisco __49__ into a city overnight because of the nearby discovery of gold in 1848. A __50__ rush to California took place. Wagon trains plodded their __51__ way across 2,000 miles of prairie and mountains, while hundreds of sailing vessels made the __52__ hazardous trip around the Horn. The vessels disgorged thousands of passengers, then the crews __53__ their ships and hundreds of vessels were left to rot in the bay. Within two years, California had enough population to become a __54__ and San Francisco was for many years the hub of that newly-arrived population.

The city's present __55__ is due to an excellent climate, an easy style of living, good food, and numerous tourist attractions. The city is famous for its cable cars which "clang and bang" up the __56__ hills, and for its excellent seafood stalls along the wharf. Most visitors arriving from nations in the Pacific Basin spend several days getting to know the town.

注意：此部分试题请在**答题卡 2** 上作答

A) pollution	I) equally
B) sprang	J) shrank
C) hopefully	K) small
D) state	L) possession
E) deserted	M) dangerous
F) seized	N) flat
G) steep	O) great
H) popularity	

Section B

Directions: *There are 2 passages in this section. Each passage is followed by some questions or unfinished statements. For each of them there are four choices marked A), B), C) and D). You should decide on the best choice and mark the corresponding letter on the **Answer Sheet 2** with a single line through the centre.*

Passage one

Questions 57 to 61 are based on the following passage.

Do you find getting up in the morning so difficult that it's painful? This might be called laziness, but Dr. Kleitman has a new explanation. He has proved that everyone has a daily energy cycle.

During the hours when you labour through your work you may say that you're "hot". That's true. The time of day when you feel most energetic is when your cycle of body temperature is at its peak. For some people the peak comes during the forenoon. For others it comes in the afternoon or evening. No one has discovered why this is so, but it leads to such familiar monologues as "Get up, John! You'll be late for work again ! " The possible explanation to the trouble is that John is at his temperature-and-energy peak in the evening. Much family quarrelling ends when husbands and wives realize what these energy cycles mean, and which cycle each member of the family has.

You can't change your energy cycle, but you can learn to make your life fit it better. Habit can help, Dr. Kleitman believes. Maybe you're sleepy in the evening but feel you must stay up late anyway. Counteract your cycle to some extent by habitually staying up later than you want to. If your energy is low in the morning, but you have an important job to do early in the day, rise before your usual hour. This won't change your cycle, but you'll get up steam and work better at your low point.

Get off to a slow start which saves your energy. Get up with a leisurely yawn and stretch. Sit on the edge of the bed a minute before putting your feet on the floor. Avoid the troublesome search for clean clothes by laying them out the night before. Whenever possible, do routine work in the afternoon and save tasks requiring more energy or concentration for your sharper hours.

注意：此部分试题请在答题卡 2 上作答

57. If a person finds getting up early a problem, most probably _____.

    A) he is very lazy

    B) he refuses to follow his own energy cycle

    C) he is not sure whether his energy is low

    D) he is at his energy peak in the afternoon or evening

58. Which of the following may lead to family quarrels according to the passage?

    A) Unawareness of energy cycles.

    B) Familiar monologues.

    C) A change in a family member's energy cycle.

    D) Attempts to control the energy cycle of other family members.

59. If one wants to work more efficiently at his low point in the morning, he should_____.

    A) change his energy cycle   B) overcome his laziness

    C) get up earlier than usual   D) go to bed later

60. You are advised to rise with a yawn and stretch because it will_____.

    A) help to keep your energy for the day's work

    B) help you to control your temper all the day

    C) enable you to concentrate on your routine work

    D) keep your energy cycle under control all day

61. Which of the following statements is NOT TRUE?

    A) Getting off to work with a minimum effort helps save one's energy.

    B) Dr. Kletman explains why people reach their peaks at different hours of day.

    C) Children have energy cycles, too.

    D) Habit helps a person adapt to his own energy cycle.

Passage Two

Questions 62 to 66 are based on the following passage.

    Reading to oneself is a modern activity that was almost unknown to the scholars of the classical and medieval worlds, while during the fifteenth century the term "reading" undoubtedly meant reading aloud. Only during nineteenth century did silent reading become commonplace.

    One should be wary, however, of assuming that silent reading came about simply because reading aloud is distraction to others. Examination of factors related to the historical development of silent reading reveals that it became the usual mode of reading for most adult reading tasks mainly because the tasks themselves changed in character.

The last century saw a steady gradual increase in literacy, and thus in the number of readers. As readers increased, so the number of potential listeners declined, and thus there was some reduction in the need to read aloud. As reading for the benefit of listeners grew less common, so came the flourishing of reading as a private activity in such public places as libraries, railway carriages and offices, where reading aloud would cause distraction to other readers.

Towards the end of the century there was still considerable argument over whether books should be used for information or treated respectfully, and over whether the reading of material such as newspapers was in some way mentally weakening. Indeed this argument remains with us still in education. However, whatever its virtues, the old shared literacy culture had gone and was replaced by printed mass media on the one hand and by books and periodicals for a specialized readership on the other.

By the end of the century students were being recommended to adopt attitudes to books and to use skills in reading them which were inappropriate, if not impossible, for the oral reader. The social, cultural, and technological changes in the century had greatly altered what the term "reading" implied.

注意： 此部分试题请在**答题卡 2** 上作答

62. Why was reading aloud common before the nineteenth century?

A) Reading aloud had been very popular.

B) There were few places available for private reading.

C) Few people could read for themselves.

D) People relied on reading for enjoyment.

63. The development of silent reading during the nineteenth century indicated_____.

A) a change in the status of literate people

B) a change in the nature of reading

C) an increase in the average age of readers

D) an increase in the number of books

64. Educationalists are still arguing about_____.

A) the importance of silent reading

B) the amount of information obtained by books and newspapers

C) the effects of reading aloud and silent reading

D) The value of different types of reading material

65. The emergence of the mass media and of specialized periodicals showed that_____.

A) standards of literacy had declined

B) reader's interests had diversified

C) printing techniques had improved

D) educationalist's attitudes had changed

66. What is the writer's purpose in writing this passage?

A) To explain how present-day reading habits developed.

B) To change people's attitudes to reading.

C) To show how reading methods have improved.

D) To encourage the growth of reading.

Part V   Error Correction   (15 minutes)

Directions: *The passage contains TEN errors. Each indicated line contains a maximum of ONE error. In each case, only ONE word involved. You should proofread the passage and correct it in the following way.*

For a wrong word, underline the wrong word and write the correct one in the blank provided at the end of the line.

For a missing word, mark the position of the missing word with a " ^ "sign and write the word you believe to be missing in the blank provided at the end of the line.

For an unnecessary word, cross the unnecessary word with a slash " / " and put the word in the blank provided at the end of the line.

EXAMPLE

When ^art museum wants a new exhibit,   (1)_____an_____

it never buys things in finished form and hangs   (2)____never____

them on the wall. When a natural history museum

wants an exhibition, it must often build it.   (3)____exhibit____

注意：此部分试题在**答题卡2**上，请在**答题卡2**上作答

Part VI   Translation   (5 minutes)

Directions: *Complete the sentences on **Answer Sheet 2** by translating the Chinese given in brackets into English.*

注意：此部分试题在**答题卡2**上，请在**答题卡2**上作答

Part I　　　　　　　　**Writing**　　　　　　　　**(30 minutes)**

Directions: *For this part, you are allowed thirty minutes to write a composition on the topic **My View on Going after Fashion**. You should write no less than 120 words and you should base your composition on the outline (given in Chinese) below:*

1. 一些学生喜欢赶时髦；
2. 另一些学生喜欢追求个性；
3. 我的看法。

My View on Going after Fashion

答题卡 1 （Answer Sheet 1）

_____
_____
_____
_____
_____
_____
_____
_____
_____
_____
_____
_____

Part Ⅱ   **Reading Comprehension (Skimming and Scanning)**   **(15 minutes)**

1. [Y] [N] [NG]   2. [Y] [N] [NG]   3. [Y] [N] [NG]   4. [Y] [N] [NG]
5. [Y] [N] [NG]   6. [Y] [N] [NG]   7. [Y] [N] [NG]

必须使用黑色签字笔书写，在答题区域内作答，超出以下矩形边框限定区域的答案无效。

8. Owing to the variety of immigrant groups, New York is often referred to as a true _____.

9. The length of New York City is _____.

10. The bus system is like the subway in that it has its own _____.

答题卡 2（Answer Sheet 2）

Part Ⅲ

Section A

11. [A] [B] [C] [D]   16. [A] [B] [C] [D]   21. [A] [B] [C] [D]

12. [A] [B] [C] [D]   17. [A] [B] [C] [D]   22. [A] [B] [C] [D]

13. [A] [B] [C] [D]   18. [A] [B] [C] [D]   23. [A] [B] [C] [D]

14. [A] [B] [C] [D]   19. [A] [B] [C] [D]   24. [A] [B] [C] [D]

15. [A] [B] [C] [D]   20. [A] [B] [C] [D]   25. [A] [B] [C] [D]

Section B

26. [A] [B] [C] [D]   31. [A] [B] [C] [D]

27. [A] [B] [C] [D]   32. [A] [B] [C] [D]

28. [A] [B] [C] [D]   33. [A] [B] [C] [D]

29. [A] [B] [C] [D]   34. [A] [B] [C] [D]

30. [A] [B] [C] [D]   35. [A] [B] [C] [D]

Part Ⅲ

Section C

必须使用黑色签字笔书写，在答题区域内作答，超出以下矩形边框限定区域的答案无效。

The lack of (36)_____is the reason most people fail in attaining their goals. Many organizational analysts and (37) _____ consultants consider persistence to be the (38) _____ _____ key to success at both the organizational and (39) _____ level.

Success (40) _____ comes easily on the first try. What (41) _____ the successful from the unsuccessful is persistence. Successful people also fail (42) _____ but they do not let their failures (43) _____ their spirit. (44) _____ _____. And again and again. Until they succeed. (45)_____, usually passing the blame on to someone or something else, and learn nothing from their experience other than perfecting their scapegoat techniques. Successful people expect periodic defeats, learn what went wrong and why, don't waste time looking for someone to blame, make necessary adjustments, and try again. (46) _____. If you are not persistent, you will almost certainly fail.

Part IV　　　**Section A**　　　　　　　**Section B**

47. [A][B][C][D][E][F][G][H][I][J][K][L][M][N][O]　　57. [A][B][C][D]

48. [A][B][C][D][E][F][G][H][I][J][K][L][M][N][O]　　58. [A][B][C][D]

49. [A][B][C][D][E][F][G][H][I][J][K][L][M][N][O]　　59. [A][B][C][D]

50. [A][B][C][D][E][F][G][H][I][J][K][L][M][N][O]　　60. [A][B][C][D]

51. [A][B][C][D][E][F][G][H][I][J][K][L][M][N][O]　　61. [A][B][C][D]

52. [A][B][C][D][E][F][G][H][I][J][K][L][M][N][O]　　62. [A][B][C][D]

53. [A][B][C][D][E][F][G][H][I][J][K][L][M][N][O]　　63. [A][B][C][D]

54. [A][B][C][D][E][F][G][H][I][J][K][L][M][N][O]　　64. [A][B][C][D]

55. [A][B][C][D][E][F][G][H][I][J][K][L][M][N][O]　　65. [A][B][C][D]

56. [A][B][C][D][E][F][G][H][I][J][K][L][M][N][O]　　66. [A][B][C][D]

Part V

　　Until the very latest moment of his existence, man has been bound to the planet on which he originated and developed. Now he had the capability to leave that planet and move out into the universe to those worlds which he has known previously only directly. MEN have explored parts of moon, put spaceships in orbit around another planet and possibly within the decade will land into another planet and explore it. Can we be too bold as to suggest that we may be able to colonize other planet within the not-too-distant future? Some have advocated such a procedure as a solution to the population problem: ship the excess people off to the moon. But we must keep in head the billions of dollars we might spend in carrying out the project. To maintain the earth's population at its present level, we would have to blast off into space 7,500 people every hour of every day of the year.

67＿＿＿＿＿＿

68＿＿＿＿＿＿

69＿＿＿＿＿＿

70＿＿＿＿＿＿

71＿＿＿＿＿＿

72＿＿＿＿＿＿

　　Why are we spending so little money on space exploration? Consider the great need for improving many aspects of the global environment, one is surely justified in his concern for the money and resources that they are poured into the space exploration efforts. But perhaps we should look at both sides of the coin before arriving hasty conclusions.

73＿＿＿＿＿＿

74＿＿＿＿＿＿

75＿＿＿＿＿＿

76＿＿＿＿＿＿

Part Ⅵ　　　　　　　Translation　　　　　　　(5 minutes)

必须使用黑色签字笔书写，在答题区域内作答，超出以下矩形边框限定区域的答案无效。

77. It is simple. The more preparation you do now, _____(考试前你就越不会紧张).

78. She often complains about not _____(感到自己在工作上不受赏识).

79. The results of the exam will be _____(推迟到) Friday afternoon.

80. It turns out that the price _____（开始回落）.

81. _____(据说) a foreign teacher will come to our class.

Model Test Four

Part I   **Writing**   **(30 minutes)**

注意：此部分试题在**答题卡** 1 上

Part II   **Reading Comprehension (Skimming and Scanning)**   **(15 minutes)**

*Directions: In this part, you will have 15 minutes to go over the passage quickly and answer the questions on **Answer Sheet 1**.*

For questions 1～7, mark

   Y *(for **YES**)*   *if the statement agrees with the information given in the passage;*
   N *(for **NO**)*   *if the statement contradicts the information given in the passage;*
   NG *(for **NOT GIVEN**)*   *if the information is not given in the passage.*

For questions 8～10, complete the sentence with the information given in the passage.

Policy on Student Privacy Rights Policy Statement

Under the Family Educational Rights and Privacy Act (FERPA), you have the right to:

● inspect and review your education records;

● request an amendment to your education records if you believe they are inaccurate or misleading;

● request a hearing if your request for an amendment is not resolved to your satisfaction;

● consent to disclosure of personally identifiable information from your education records, except to the extent that FERPA authorizes disclosure without your consent;

● file a complaint with the U.S. Department of Education Family Policy Compliance Office if you believe your rights under FERPA have been violated.

1. Inspection

What are education records?

Education records are records maintained by the university that are directly related to student. These include biographic and demographic data, application materials, course schedules, grades and work-study records. The term does not include:

● information contained in the private files of instructors and administrators, used only as a personal memory aid not accessible or revealed to any other person except a temporary substitute for the maker of the record;

● Campus Police records;

- employment records other than work-study records;
- medical and psychological records used solely for treatment purposes;
- records that only contain information about individuals after they have left the university;
- any other records that do not meet the above definition of education records.

How do I inspect my education records?

- Complete an Education Inspection and Review Request Form (available online as a PDF document or for The HUB, 12C Warner Hall) and return it to The HUB.
- The custodian of the education record you wish to inspect will contact you to arrange a mutually convenient time for inspection, not more than 45 days after you request. The custodian or designee will be present during your inspection.
- You will not be permitted to review financial information, including your parents' financial information; or confidential letters of recommendation, if you have waived your right to inspect such letters.
- You can get copies of your education records from the office where they are kept for 25 cents per page, prepaid.

2. Amendment
How do I amend my educational records?

- Send a written, signed request for amendment to the Vice President for Enrollment, Carnegie Mellon University, 610 Warner Hall, Pittsburgh, PA 15213. Your request should specify the record you want to have amended and the reason for amendment.
- The university will reply to you no later than 45 days after your request. If the university does not agree to amend the record, you have a right to a hearing on the issue.

3. Hearing
How do I request a hearing?

Send a written, signed request for a hearing to the Vice President for Enrollment , Carnegie Mellon University, 610 Warner Hall, Pittsburgh, PA 15213. The university will schedule a hearing no later than 45 days after your request.

How will the hearing be conducted?

- A university officer appointed by the President for Enrollment, who is not affiliated with your enrolled college, will conduct the hearing.
- You can bring others, including an attorney, to the hearing to assist or represent you. If your attorney will be present, you must notify the university ten days in advance of the hearing so

that the university can arrange to have an attorney present too, if desired.

- The university will inform you of its decision, in writing, including a summary of the evidence presented and the reasons for its decision, no later than 45 days after the hearing.
- If the university decides not to amend the record, you have a right to add a statement to the record that explains your side of the story.

4. Disclosure

Carnegie Mellon generally will not disclose personally identifiable information from your education records without your consent except for directory information and other exceptions specified by law.

What is directory information?

Directory information is personally identifiable information of a general nature that may be disclosed without your consent, unless you specifically request the university not to do so. It is used for purpose like compiling campus directories.

If you do not want your directory information to be disclosed, you must notify The HUB, 12C Warner Hall, in writing within the first 15 days of the semester.

Notifying The HUB covers only the disclosure of centralized records. Members of individual organizations such as fraternities, sororities, athletics, etc. must also notify those organizations to restrict the disclosure of directory information.

Carnegie Mellon has defined directory information as the following:
- your full name
- local/campus address
- local/campus telephone number
- email user id and address
- major, department, college
- class status (freshman, sophomore, junior, senior, undergraduate, or graduate)
- dates of attendance (semester begin and end dates)
- enrollment status (full, half, of part time)
- date(s) of graduation
- degrees awarded
- sorority or fraternity affiliation

For students participating in intercollegiate athletics, directory information also includes:
- height, weight
- sport of participation

What are the other exceptions?

Under FERPA, Carnegie Mellon may release personally identifiable information from your education records without your prior consent to:

- school officials with legitimate educational interests;
- certain federal officials in connection with federal program requirements;
- organizations involved in awarding financial aid;
- state and local officials who are legally entitled to the information;
- testing agencies such as the Educational Testing Service, for the purpose of developing, validating, researching and administering tests;
- accrediting agencies, in connection with their accrediting functions;
- parents of dependent students (as defined in section 152 of the Internal Revenue Service Code);
- appropriate parties in a health or safety emergency, if necessary to protect the health or safety of the student or other individuals;
- officials of another school in which the student seeks or intends to enroll;
- victims of violent crimes or non-forcible sexual offenses (the results of final student disciplinary proceedings);
- parents or legal guardians of student under 21 years of age (information regarding violations of university drug and alcohol polices);
- Courts (records relevant to legal actions initiated by students, parents or the university).

5. Complaints

If you believe the university has not complied with FERPA, you can file a complaint with the:

Family Policy Compliance Office

Department of Education

400 Maryland Avenue, S.W.

Washington, DC20202-4605

注意：此部分试题请在**答题卡1**上作答

1. This article has university students as its target audience.
2. Under FERPA, students are entitled to request an amendment to their education records whenever they wish.
3. The education records are kept by the students themselves.
4. A student's demand for the inspection of his records must be met within a month.
5. When the request for an amendment is refused by the university, the student may ask for a hearing.
6. The university is free to disclose the directory information of students without their consent.
7. In a hearing for the amendment of education records, both the student and the university may hire attorneys.

8~10题请见**答题卡1**

Part III   Listening Comprehension   (35 minutes)

Section A

Directions: *In this section, you will hear 8 short and 2 long conversations. At the end of each conversation, one or more questions will be asked about what was said. Both the conversation and the question will be spoken only once. After each question there will be a pause. During the pause, you must read the four choices marked A), B), C)and D), and decide which is the best answer. Then mark the corresponding letter on the **Answer Sheet** 2 with a single line through the center.*

注意：此部分试题请在**答题卡 2** 上作答

11. A) He'll be speaking at the end of the meeting.

    B) He was supposed to speak last night instead.

    C) He suddenly decided not to speak.

    D) He already spoke very briefly tonight.

12. A) The man shouldn't expect her to go along.

    B) She doesn't think she has enough money.

    C) She'll go even though the movie is bad.

    D) The man should count the number of people going.

13. A) Both the man and the woman have no time to look at the gift.

    B) The man can't imagine what his friends get for him.

    C) The man already knows what Betty will say.

    D) The man is anxious to see Betty's reaction to the gift.

14. A) She wasn't really studying.

    B) She hadn't finished writing her articles.

    C) She had furnished her house.

    D) She could write beautifully.

15. A) The problem may have been a very complicated one.

    B) No one can do it.

    C) The woman thinks that the problem is too easy.

    D) The man can solve the problem himself.

16. A) The janitor is too busy to do his work.

    B) The sanitary conditions of an apartment.

    C) The relationship between the janitor and the two speakers.

    D) The architecture of a building.

17. A) He can't tear either piece of cloth.

    B) He wants part of each piece of cloth.

    C) The pieces of cloth are made by a secret process.

    D) The pieces of cloth seem identical to him.

18. A) Look around before going home.

    B) Prefer to argue about it.

    C) Disagree with the woman.

    D) Apologize to the woman.

Questions 19 to 22 are based on the conversation you have just heard.

19. A) Taking an exam.   B) Drawing graphs.

    C) Giving presentation.   D) Having a class discussion.

20. A) Business.   B) Fashion design.

    C) Chemistry.   D) Art appreciation.

21. A) The man came here on feet.   B) The man drove here.

    C) The man came here by bus.   D) The man came here by taxi.

22. A) If anything can go wrong, it will.   B) Where there is a will, there is a way.

    C) God helps those who help themselves.   D) A friend in need is a friend indeed.

Questions 23 to 25 are based on the conversation you have just heard.

23. A) An exhibition of paintings.   B) An opera.

    C) A modern dance production.   D) A Broadway play.

24. A) Artists.   B) Musicians.

    C) Salesmen.   D) Tour Guides.

25. A) As interesting.   B) As cold.

    C) As popular.   D) As huge.

Section B

Directions: *In this section, you will hear 3 short passages. At the end of each passage, you will hear some questions. Both the passage and the questions will be spoken only once. After you hear a question, you must choose the best answer from the four choices marked A), B), C) and D). Then mark the corresponding letter on the **Answer Sheet 2** with a single line through the center.*

注意：此部分试题请在**答题卡 2** 上作答

Passage One

Questions 26 to 29 are based on the passage you have just heard.

26. A) To keep fish alive.   B) To punish criminals.

    C) To preserve dead bodies.   D) To help heal wounds.

27. A) For making salted fish.   B) For stealing salt.
    C) For taking salt from the king's table. D) For selling salt.

28. A) He would lose his life.   B) He would lose an ear.
    C) He would lose all his salt.   D) He would be heavily fined.

29. A) Three thousand years ago.   B) When man began to salt fish.
    C) When man began to preserve the dead. D) No one knows.

Passage Two

Questions 30 to 32 are based on the passage you have just heard.

30. A) A few inches above the knee.   B) A little below the knee.
    C) Down to the ankle.   D) Floor-length.

31. A) Boots.   B) Sneakers.
    C) Slippers.   D) Leather shoes.

32. A) Fashions change overtime.

    B) Men are thriftier than women.

    C) Skirts and shoes are more important than other clothing.

    D) Some clothing may suit all occasions.

Passage Three

Questions 33 to 35 are based on the passage you have just heard.

33. A) Vacations.   B) Wages.
    C) Overcrowded classrooms.   D) Paid sick leaves.

34. A) They want the teachers to resign.

    B) They want the teachers to return to work.

    C) They are very sympathetic toward the strike.

    D) They are refusing to comment on the situation.

35. A) Parent Board.   B) District Court.
    C) Teachers' Union.   D) School Committee.

Section C

Directions: *In this section, you will hear a passage three times. When the passage is read for the first time, you should listen carefully for its general idea. When the passage is read for the second time, you are required to fill in the blanks numbered from 36 to 43 with the exact words you have just heard. For blanks numbered from 44 to 46 you are required to fill in the missing information. You can either use the exact words you have just heard or write down the main points in your own words. Finally, when the passage is read for the third time, you should check what you have written.*

注意：此部分试题在答题卡 2 上，请在答题卡 2 上作答

Part IV   Reading Comprehension (Reading in Depth)   (25 minutes)

Section A

Directions: *In this section, there is a passage with ten blanks. You are required to select one word for each blank from a list of choices given in a word bank following the passage. Read the passage through carefully before making your choices. Each choice in the bank is identified by a letter. Please mark the corresponding letter for each item on **Answer Sheet 2** with a single line through the centre.* <u>*You may not use any of the words in the bank*</u> <u>*more than once.*</u>

Questions 47 to 56 are based on the following passage.

Soon after an event takes place, newspapers are on the streets to give the details. Wherever anything happens in the world,　47　are on the spot to gather the news. Newspapers have one　48　purpose, to get the news as quickly as possible from its source, from those who make it to those who want to　49　it.

Radio, telegraph, television, and other　50　brought competition for newspapers. So did the development of magazines and other means of communication. However, this competition merely spurred the newspapers on. They quickly make use of the newer and faster means of communication to　51　the speed and thus the efficiency of their own operations. Today more newspapers are printed and read than ever before. Competition also led newspapers to branch out into many other　52　. Besides keeping readers informed of the latest news, today's newspapers educate and influence readers about politics and other important and　53　matters. Newspapers influence readers'　54　choices through advertising. Most newspapers depend　55　on advertising for their very existence.

Newspapers are sold at a price that fails to cover even a　56　fraction of the cost of production. The main source of income for most newspapers is commercial advertising. The success in selling advertising depends on a newspaper's value to advertisers. This is measured in terms of circulation.

注意：此部分试题请在**答题卡 2** 上作答

A) inventions		I) basic	
B) funny		J) events	
C) know		K) economic	
D) heavily		L) slightly	
E) fields		M) reporters	
F) slow		N) produce	
G) improve		O) small	
H) serious			

Section B

Directions: *There are 2 passages in this section. Each passage is followed by some questions or unfinished statements. For each of them there are four choices marked A), B), C) and D). You should decide on the best choice and mark the corresponding letter on the **Answer Sheet 2** with a single line through the centre.*

Passage one

Questions 57 to 61 are based on the following passage.

BMW's efforts to harness the creativity of its customers began two years ago when it posted a toolkit on its website. This toolkits let BMW's customers develop ideas showing how the firm could take advantage of advances in technology and in-car online services. From the 1,000 customers who used the toolkit, BMW chose 15 and invited them to meet its engineers in Munich. Some of their ideas (which remain under wraps for now) have since reached the prototype stage, says BMW. "They were so happy to be invited by us, and that our technical experts were interested in their ideas," says Mr Reimann. "They didn't want any money."

Westwood Studios, a game developer now owned by EA, first noticed its customers innovating its products after the launch of a game "Red Alert" in 1996: gamers were making new content for existing games and posting it freely on fan websites. Westwood made a conscious decision to embrace this phenomenon. Soon it was shipping basic game-development tools with its games, and by 1999 had a dedicated department to feed designers and producers working on new projects with customer innovations of existing ones. "The fan community has had a tremendous influence on game design," says Mr Verdu, "and the games are better as a result."

Researchers call such customers "lead users". GE's healthcare division calls them "luminaries". They tend to be well-published doctors and research scientists from leading medical institutions, says GE, which brings up to 25 luminaries together at regular medical advisory board sessions to discuss the evolution of GE's technology. GE then shares some of its advanced technology with a subset of luminaries who are from an "inner sanctum of good friends", says Sholom Ackelsberg of GE Healthcare. GE's products then emerge from collaboration with these groups.

注意：此部分试题请在答题卡2上作答

57. Why does BMW post a toolkit on its website?

   A) Because it wants to interest more customers.

   B) Because it wants to improve their website.

   C) Because it wants their customers to give advices or ideas on their products.

   D) Because it wants see if the customers' ideas match their prototype.

58. We may conclude from the text that _____.

   A) EA is a computer game producer

   B) EA is the largest hi-tech company in the world

C) "Red Alert" made its first appearance before 1996

D) Westwood Studios used to be owned by EA for many years

59. Which of the following behavior does not reflect that we are now in a customer-driven market ?

A) BMW posts a toolkit to collect customers' ideas.

B) GE brings up 25 luminaries to discuss the evolution of GE's technology.

C) Westwood establishes a department to deal with customers' innovations.

D) GE's healthcare division calls some of the well-published doctors and research scientists "luminaries".

60. Which of the following can replace the word "customer- driven"?

A) customer-centered   B) customer-satisfied

C) customer-analyzed   D) customer-evaluate

61. Customers invited by BMW didn't want any money, instead , they just want _____.

A) to be invited in a conference

B) their suggestions and ideas to be accepted by the company and be of use in the cars' upgrade

C) take a look at BMW's newest models

D) get together and exchange experience on driving the BMWs

Passage Two

Questions 62 to 66 are based on the following passage.

Where do pesticides fit into the picture of environmental disease? We have seen that they now pollute soil, water, and food, that they have the power to make our streams fishless and our gardens and woodlands silent and birdless. Man, however much he may like to pretend the contrary, is part of nature. Can he escape a pollution that is now so thoroughly distributed throughout our world ?

We know that even single exposures to these chemicals, if the amount is large enough, can cause extremely severe poisoning. But this is not the major problem. The sudden illness or death of farmers, farm workers, and others exposed to sufficient quantities of pesticides are very sad and should not occur. For the population as a whole, we must be more concerned with the delayed effects of absorbing small amounts of the pesticides that invisibly pollute our world.

Responsible public health officials have pointed out that the biological effects of chemicals are cumulative over long periods of time, and that the danger to the individual may depend on the sum of the exposures received throughout his lifetime. For these very reasons the danger is easily ignored. It is human nature to shake off what may seem to us a threat of future disaster. "Men are naturally most impressed by diseases which have obvious signs", says a wise physician, Dr. Rene Dubos, "yet some of their worst enemies slowly approach them unnoticed."

注意：此部分试题请在答题卡 2 上作答

62. Which of the following is closest in meaning to the sentence "Man, however much he may like to pretend the contrary, is part of nature."?

A) Man appears indifferent to what happens in nature.

B) Man can escape his responsibilities for environmental protection.

C) Man acts as if he does not belong to nature.

D) Man can avoid the effects of environmental pollution.

63. What is the author's attitude towards the environmental effects of pesticides?

    A) Indifferent   B) Pessimistic

    C) Concerned   D) Defensive

64. In the author's view, the sudden death caused by exposure to large amounts of pesticides _____.

    A) now occurs most frequently among all accidental deaths

    B) is not the worst of the negative consequences resulting from the use of pesticides

    C) has sharply increased so as to become the center of public attention

    D) is unavoidable because people can't do without pesticides in farming

65. People tend to ignore the delayed effects of exposure to chemicals because _____.

    A) limited exposure to them doesn't do much harm to people's health

    B) the present is more important for them than the future

    C) the danger does not become apparent immediately

    D) humans are able to withstand small amounts of poisoning

66. It can be concluded from Dr Dubo's remarks that _____.

    A) people find invisible diseases difficult to deal with

    B) people tend to overlook hidden dangers caused by pesticides

    C) diseases with obvious signs are easy to cure

    D) attacks by hidden enemies tend to be fatal

Part V   Cloze   (15 minutes)

Directions: *There are 20 blanks in the following passage. For each blank there are four choices marked A), B), C) and D) on the right side of the paper. You should choose the ONE that best fits into the passage. Then mark the corresponding letter on Answer Sheet 2 with a single line through the center.*

注意：此部分试题请在**答题卡 2**上作答

Today the car is the most popular sort of transportation in all of the United States. It has completely __67__ the horse as a __68__ of

                      67. A. denied   B. reproduced
                                      C. replaced   D. ridiculed

                      68. A. means   B. mean
everyday transportation. Americans use their car   C. types   D. kinds
for __69__ 90% of all __70__ business. Most   69. A. hardly   B. nearly
                                      C. certainly   D. somehow

                      70. A. personal   B. personnel
                                      C. manual   D. artificial
Americans are able to __71__ cars. The average   71. A. buy   B. sell

price of a ___72___ made car was 50 in 1950, 470

in 1960 and up to 750 ___73___ 1975.

    During this period, American car manufac-
turers set about ___74___ their products and work
efficiency. As a result, the yearly income of the
___75___ family increased from 1950

to 1975 ___76___ than the price of cars. For this

reason ___77___ a new car takes a smaller ___78___ of

a family's total earnings today. In 1951 ___79___ it

took 8.1 months of an average family's ___80___ to

buy a new car. In 1962 a new car ___81___ 8.3 of a
family's annual earnings, by 1975 it only took
4.75 ___82___ income.

    In addition, the 1975 cars were technically
___83___ to models from previous years. The ___84___

of automobile extends throughout the economy
___85___ the car is so important to American.

Americans spend more money ___86___ keeping
their cars running than on any other item.

	C. race	D. see
72. A. quickly		B. regularly
	C. rapidly	D. recently
73. A. on		B. in
	C. before	D. after
74. A. raising		B. making
	C. reducing	D. improving
75. A. unusual		B. smallest
	C. average	D. biggest
76. A. slower		B. equal
	C. faster	D. less
77. A. bringing		B. obtain
	C. bought	D. purchasing
78. A. part		B. half
	C. number	D. quality
79. A. clearly		B. proportionally
	C. percentage	D. suddenly
80. A. income		B. work
	C. plans	D. debts
81. A. used		B. spent
	C. cost	D. needed
82. A. months		B. years
	C. family	D. year
83. A. famous		B. superior
	C. fastest	D. better
84. A. running		B. notice
	C. influence	D. affect
85. A. then		B. as
	C. so	D. which
86. A. to		B. in
	C. of	D. for

Part VI   Translation   (5 minutes)

Directions: *Complete the sentences on **Answer Sheet 2** by translating the Chinese given in brackets
        into English.*

注意：此部分试题在**答题卡 2** 上, 请在**答题卡 2** 上作答

答题卡 1 （Answer Sheet 1）

<table>
<tr><td>学校:</td><td colspan="2"></td><td colspan="16" style="text-align:center">准　　考　　证　　号</td></tr>
<tr><td rowspan="2">姓名:</td><td colspan="2"></td><td>[0]</td><td>[0]</td><td>[0]</td><td>[0]</td><td>[0]</td><td>[0]</td><td>[0]</td><td>[0]</td><td>[0]</td><td>[0]</td><td>[0]</td><td>[0]</td><td>[0]</td><td>[0]</td><td>[0]</td></tr>
<tr><td colspan="2"></td><td>[1]</td><td>[1]</td><td>[1]</td><td>[1]</td><td>[1]</td><td>[1]</td><td>[1]</td><td>[1]</td><td>[1]</td><td>[1]</td><td>[1]</td><td>[1]</td><td>[1]</td><td>[1]</td><td>[1]</td></tr>
<tr><td rowspan="8">填涂要求</td><td colspan="2" style="text-align:center">正确填涂</td><td>[2]</td><td>[2]</td><td>[2]</td><td>[2]</td><td>[2]</td><td>[2]</td><td>[2]</td><td>[2]</td><td>[2]</td><td>[2]</td><td>[2]</td><td>[2]</td><td>[2]</td><td>[2]</td><td>[2]</td></tr>
<tr><td colspan="2" style="text-align:center">■</td><td>[3]</td><td>[3]</td><td>[3]</td><td>[3]</td><td>[3]</td><td>[3]</td><td>[3]</td><td>[3]</td><td>[3]</td><td>[3]</td><td>[3]</td><td>[3]</td><td>[3]</td><td>[3]</td><td>[3]</td></tr>
<tr><td colspan="2"></td><td>[4]</td><td>[4]</td><td>[4]</td><td>[4]</td><td>[4]</td><td>[4]</td><td>[4]</td><td>[4]</td><td>[4]</td><td>[4]</td><td>[4]</td><td>[4]</td><td>[4]</td><td>[4]</td><td>[4]</td></tr>
<tr><td colspan="2"></td><td>[5]</td><td>[5]</td><td>[5]</td><td>[5]</td><td>[5]</td><td>[5]</td><td>[5]</td><td>[5]</td><td>[5]</td><td>[5]</td><td>[5]</td><td>[5]</td><td>[5]</td><td>[5]</td><td>[5]</td></tr>
<tr><td colspan="2" style="text-align:center">错误填涂</td><td>[6]</td><td>[6]</td><td>[6]</td><td>[6]</td><td>[6]</td><td>[6]</td><td>[6]</td><td>[6]</td><td>[6]</td><td>[6]</td><td>[6]</td><td>[6]</td><td>[6]</td><td>[6]</td><td>[6]</td></tr>
<tr><td colspan="2" style="text-align:center">☑ ☒ ⊘</td><td>[7]</td><td>[7]</td><td>[7]</td><td>[7]</td><td>[7]</td><td>[7]</td><td>[7]</td><td>[7]</td><td>[7]</td><td>[7]</td><td>[7]</td><td>[7]</td><td>[7]</td><td>[7]</td><td>[7]</td></tr>
<tr><td colspan="2" style="text-align:center">⊙ ● ▰</td><td>[8]</td><td>[8]</td><td>[8]</td><td>[8]</td><td>[8]</td><td>[8]</td><td>[8]</td><td>[8]</td><td>[8]</td><td>[8]</td><td>[8]</td><td>[8]</td><td>[8]</td><td>[8]</td><td>[8]</td></tr>
<tr><td colspan="2"></td><td>[9]</td><td>[9]</td><td>[9]</td><td>[9]</td><td>[9]</td><td>[9]</td><td>[9]</td><td>[9]</td><td>[9]</td><td>[9]</td><td>[9]</td><td>[9]</td><td>[9]</td><td>[9]</td><td>[9]</td></tr>
</table>

Part I　　　　　　　　　　**Writing**　　　　　　　　**(30 minutes)**

Directions: *For this part, you are allowed thirty minutes to write a composition on the topic* **Wealth and Happiness.** *You should write no less than 120 words and you should base your composition on the outline (given in Chinese) below:*

1. 某些人渴求财富，并为此奋斗;
2. 有时财富的确能带来幸福;
3. 但也有例外，如……

Wealth and Happiness

_____
_____
_____
_____
_____
_____

答题卡 1 （Answer Sheet 1）

_____
_____
_____
_____
_____
_____
_____
_____
_____
_____
_____

Part II   Reading Comprehension (Skimming and Scanning)   (15 minutes)

1. [Y] [N] [NG]   2. [Y] [N] [NG]   3. [Y] [N] [NG]   4. [Y] [N] [NG]

5. [Y] [N] [NG]   6. [Y] [N] [NG]   7. [Y] [N] [NG]

必须使用黑色签字笔书写，在答题区域内作答，超出以下矩形边框限定区域的答案无效。

8. If a student feels that his right under FERPA has been violated, he could file a _____ with the U.S. Department of Education Family Policy Compliance Office.

9. Student who wants to inspect his education records must fill out an _____.

10. The directory information for intercollegiate athletes include additional information such as_____.

答题卡 2（Answer Sheet 2）

学校:	准 考 证 号

<table>
<tr><td>学校:</td><td colspan="15" align="center">准　考　证　号</td></tr>
</table>

姓名:	[0]	[0]	[0]	[0]	[0]	[0]	[0]	[0]	[0]	[0]	[0]	[0]	[0]	[0]	[0]
	[1]	[1]	[1]	[1]	[1]	[1]	[1]	[1]	[1]	[1]	[1]	[1]	[1]	[1]	[1]
	[2]	[2]	[2]	[2]	[2]	[2]	[2]	[2]	[2]	[2]	[2]	[2]	[2]	[2]	[2]
填	[3]	[3]	[3]	[3]	[3]	[3]	[3]	[3]	[3]	[3]	[3]	[3]	[3]	[3]	[3]
涂	[4]	[4]	[4]	[4]	[4]	[4]	[4]	[4]	[4]	[4]	[4]	[4]	[4]	[4]	[4]
要	[5]	[5]	[5]	[5]	[5]	[5]	[5]	[5]	[5]	[5]	[5]	[5]	[5]	[5]	[5]
求	[6]	[6]	[6]	[6]	[6]	[6]	[6]	[6]	[6]	[6]	[6]	[6]	[6]	[6]	[6]
	[7]	[7]	[7]	[7]	[7]	[7]	[7]	[7]	[7]	[7]	[7]	[7]	[7]	[7]	[7]
	[8]	[8]	[8]	[8]	[8]	[8]	[8]	[8]	[8]	[8]	[8]	[8]	[8]	[8]	[8]
	[9]	[9]	[9]	[9]	[9]	[9]	[9]	[9]	[9]	[9]	[9]	[9]	[9]	[9]	[9]

正确填涂 ■

错误填涂 ☑ ☒ ⧄ ⬭ ⬤ ▣

Part III

Section A

11. [A] [B] [C] [D]
12. [A] [B] [C] [D]
13. [A] [B] [C] [D]
14. [A] [B] [C] [D]
15. [A] [B] [C] [D]

16. [A] [B] [C] [D]
17. [A] [B] [C] [D]
18. [A] [B] [C] [D]
19. [A] [B] [C] [D]
20. [A] [B] [C] [D]

21. [A] [B] [C] [D]
22. [A] [B] [C] [D]
23. [A] [B] [C] [D]
24. [A] [B] [C] [D]
25. [A] [B] [C] [D]

Section B

26. [A] [B] [C] [D]
27. [A] [B] [C] [D]
28. [A] [B] [C] [D]
29. [A] [B] [C] [D]
30. [A] [B] [C] [D]

31. [A] [B] [C] [D]
32. [A] [B] [C] [D]
33. [A] [B] [C] [D]
34. [A] [B] [C] [D]
35. [A] [B] [C] [D]

Part III

Section C

必须使用黑色签字笔书写，在答题区域内作答，超出以下矩形边框限定区域的答案无效。

My cup of tea (36) _____ to the sort of thing that pleases or (37) _____ to me. The (38) _____ is nearly always used (39) _____. The (40) _____ came into use between the First and Second World Wars. In the Victorian age the (41) _____ of tea by all classes had not yet, (42). _____ among men, become common. A more likely metaphor then, (43) _____ from food or drink for something not to one's taste, would have been, say, "not my pot of beer", or among the well-to-do classes, "not my glass of wine". (44) _____

_____

Later, tea would come to be regarded as a universal social drink. (45) _____

_____

This variation would naturally lend itself to the expression: (46) _____

_____: e.g. an entertainment at a theatre, a book, etc, with the meaning "Whatever others may like, that is not the sort of thing to appeal to me".

答题卡 2（Answer Sheet 2）

Part Ⅳ   Section A   Section B   **Part Ⅴ**

47. [A][B][C][D][E][F][G][H][I][J][K][L][M][N][O]   57. [A][B][C][D]   67. [A][B][C][D]   77. [A][B][C][D]

48. [A][B][C][D][E][F][G][H][I][J][K][L][M][N][O]   58. [A][B][C][D]   68. [A][B][C][D]   78. [A][B][C][D]

49. [A][B][C][D][E][F][G][H][I][J][K][L][M][N][O]   59. [A][B][C][D]   69. [A][B][C][D]   79. [A][B][C][D]

50. [A][B][C][D][E][F][G][H][I][J][K][L][M][N][O]   60. [A][B][C][D]   70. [A][B][C][D]   80. [A][B][C][D]

51. [A][B][C][D][E][F][G][H][I][J][K][L][M][N][O]   61. [A][B][C][D]   71. [A][B][C][D]   81. [A][B][C][D]

52. [A][B][C][D][E][F][G][H][I][J][K][L][M][N][O]   62. [A][B][C][D]   72. [A][B][C][D]   82. [A][B][C][D]

53. [A][B][C][D][E][F][G][H][I][J][K][L][M][N][O]   63. [A][B][C][D]   73. [A][B][C][D]   83. [A][B][C][D]

54. [A][B][C][D][E][F][G][H][I][J][K][L][M][N][O]   64. [A][B][C][D]   74. [A][B][C][D]   84. [A][B][C][D]

55. [A][B][C][D][E][F][G][H][I][J][K][L][M][N][O]   65. [A][B][C][D]   75. [A][B][C][D]   85. [A][B][C][D]

56. [A][B][C][D][E][F][G][H][I][J][K][L][M][N][O]   66. [A][B][C][D]   76. [A][B][C][D]   86. [A][B][C][D]

Part Ⅵ   **Translation**   **(5 minutes)**

必须使用黑色签字笔书写，在答题区域内作答，超出以下矩形边框限定区域的答案无效。

87. The rent is reasonable and, _____ (此外)，the location is perfect.

88. Driving again after his accident must_____ (需要很大勇气).

89. Miss Smith was _____ (处于最佳状态) when she played the piano.

90. They can not choose but_____ (承认他们错了).

91. He is an artist _____ (而不是一名教师).

Model Test Five

Part I **Writing** **(30 minutes)**

注意：此部分试题在**答题卡 1** 上

Part II **Reading Comprehension(Skimming and Scanning)** **(15 minutes)**

Directions: *In this part, you will have 15 minutes to go over the passage quickly and answer the questions on* **Answer Sheet 1.**

For questions 1～7, mark

    Y *(for* **YES***)*   *if the statement agrees with the information given in the passage;*

    N *(for* **NO***)*   *If the statement contradicts the information given in the passage;*

    NG *(for* **NOT GIVEN***)*   *if the information is not given in the passage.*

For questions 8～10, complete the sentences with the information given in the passage.

The Cultural Patterning of Space

Like time, space is perceived differently in different cultures. Spatial consciousness in many Western cultures is based on a perception of objects in space, rather than of space itself. Westerners perceive shapes and dimensions, in which space is a realm of light, color, sight, and touch. Benjamin L. Whorf, in his classic work Language, Thought and Reality, offers the following explanation as one reason why Westerners perceive space in this manner. Western thought and language mainly developed from the Roman, Latin-speaking, culture, which was a practical, experience-based system. Western culture has generally followed Roman thought patterns in viewing objective "reality" as the foundation for subjective or "inner" experience. It was only when the intellectually crude Roman culture became influenced by the abstract thinking of the Greek culture that the Latin language developed a significant vocabulary of abstract, nonspatial terms. But the early Roman-Latin element of spatial consciousness, of concreteness, has been maintained in Western thought and language patterns, even though the Greek capacity for abstract thinking and expression was also inherited.

However, some cultural-linguistic systems developed in the opposite direction, that is, from an abstract and subjective vocabulary to a more concrete one. For example, Whorf tells us that in the Hopi language the word heart, a concrete term, can be shown to be a late formation from the abstract terms think or remember. Similarly, although it seems to Westerners, and especially to Americans, that objective, tangible "reality" must precede any subjective or inner experience; in fact, many Asian and other non-European cultures view inner experience as the basis for one's perceptions of physical reality. Thus although Americans are taught to perceive and react to the arrangement of objects in space and to think of space as being "wasted" unless it is filled with objects, the Japanese are trained to give meaning to space itself and to value "empty" space.

It is not only the East and the West that are different in their patterning of space. We can also see cross-cultural varieties in spatial perception when we look at arrangements of urban space in different Western cultures. For instance, in the United States, cities are usually laid out along a grid, with the axes generally north/south and east/west. Streets and buildings are numbered sequentially. This arrangement, of course, makes perfect sense to Americans. When Americans walk in a city like Paris, which is laid out with the main streets radiating from centers, they often get lost. Furthermore, streets in Paris are named, not numbered, and the names often change after a few blocks. It is amazing to Americans how anyone gets around, yet Parisians seem to do well. Edward Hall, in The Silent Language, suggests that the layout of space characteristic of French cities is only one aspect of the theme of centralization that characterizes French culture. Thus Paris is the center of France, French government and educational systems are highly centralized, and in French offices the most important person has his or her desk in the middle of the office.

Another aspect of the cultural patterning of space concerns the functions of spaces. In middle-class America, specific spaces are designated for specific activities. Any intrusion of one activity into a space that it was not designed for is immediately felt as inappropriate. In contrast, in Japan, this case is not true: Walls are movable, and rooms are used for one purpose during the day and another purpose in the evening and at night. In India there is yet another culturally patterned use of space. The function of space in India, both in public and in private places, is connected with concepts of superiority and inferiority. In Indian cities, villages, and even within the home, certain spaces are designated as polluted, or inferior, because of the activities that take place there and the kinds of people who use such spaces. Spaces in India are segregated so that high caste and low caste, males and females, secular and sacred activities are kept apart. This pattern has been used for thousands of years, as demonstrated by the archaeological evidence uncovered in ancient Indian cities. It is a remarkably persistent pattern, even in modern India, where public transportation reserves a separate space for women. For example, Chandigarh is a modern Indian city designed by a French architect. The apartments were built according to European concepts, but the Indians living there found certain aspects inconsistent with their previous use of living space. Ruth Freed, an anthropologist who worked in India, found that Indian families living in Chandigarh modified their apartments by using curtains to separate the men's and women's spaces. The families also continued to eat in the kitchen, a traditional pattern, and the living room-dining room was only used when Western guests were present. Traditional Indian village living takes place in an area surrounded by a wall. The courtyard gives privacy to each residence group. Chandigarh apartments, however, were built with large windows, reflecting the European value of light and sun, so many Chandigarh families pasted paper over the windows to recreate the privacy of the traditional courtyard. Freed suggests that these traditional

Indian patterns may represent an adaptation to a densely populated environment.

Anthropologists studying various cultures as a whole have seen a connection in the way they view both time and space. For example, as we have seen, Americans look on time without activity as "wasted" and space without objects as "wasted." Once again, the Hopi present an interesting contrast. In the English language, any noun for a location or a space may be used on its own and given its own characteristics without any reference being made to another location or space. For example, we can say in English: "The room is big" or "The north of the United States has cold winters". We do not need to indicate that "room" or "north" has a relationship to any other word of space or location. But in Hopi, locations or regions of space cannot function by themselves in a sentence. The Hopi can not say "north" by itself; they must say "in the north," "from the north," or in some other way use a directional suffix with the word north. In the same way, the Hopi language does not have a single word that can be translated as room. The Hopi word for room is a stem, a portion of a word, that means "house," "room," or "enclosed chamber," but the stem can not be used alone. It must be joined to a suffix that will make the word mean "in a house" or "from a chamber." Hollow spaces like room, chamber, or hall in Hopi are concepts that are meaningful only in relation to other spaces.

In some cultures a significant aspect of spatial perception is shown by the amount of "personal space" people need between themselves and others to feel comfortable and not crowded. North Americans, for instance, seem to require about four feet of space between themselves and people near them to feel comfortable. On the other hand, people from Arab countries and Latin America feel comfortable when they are close to each other. People from different cultures, therefore, may unconsciously infringe on each other's sense of space. Thus just as different perceptions of time may create cultural conflicts, so too may different perceptions of space.

注意：此部分试题请在答题卡 1 上作答

1. The passage is about cross-cultural spatial perceptions.

2. European cultures generally value inner personal experience more than non-European cultures do.

3. China is an example of a highly centralized society.

4. Japan and the United States are similar in that both cultures use the same space for a variety of different purposes.

5. In India, public and private space is separated for males and females.

6. The Hopi language locates places only in connection with other spaces or directions.

7. Arab, Latin American, and North American cultures all have similar perceptions of personal space.

8 ~ 10 题请见答题卡 1

Part III   Listening Comprehension   (35 minutes)

Section A

Directions: *In this section, you will hear 8 short and 2 long conversations. At the end of each conversation, one or more questions will be asked about what was said. Both the conversation and the question will be spoken only once. After each question there will be a pause. During the pause, you must read the four choices marked A), B), C) and D), and decide which is the best answer. Then mark the corresponding letter on the **Answer Sheet 2** with a single line through the center.*

注意：此部分试题请在**答题卡 2** 上作答

11. A) She can do the job.   B) She could call a friend.

     C) She's just switched off the light.   D) She's already replaced the shelf.

12. A) They want to go downtown.

     B) He wants to go to the park, but she doesn't.

     C) He doesn't know where to park the car.

     D) He wants to find out where the park is.

13. A) Company and customer.   B) Repairman and customer.

     C) Teacher and student's parent.   D) Wife and husband.

14. A) She didn't like working in a company.   B) She disliked machines.

     C) She was not good at doing business.   D) She didn't like accounting best.

15. A) He has some money to buy a new car.

     B) He fails in borrowing enough money from the woman.

     C) He will spend much money on his house.

     D) He wants to buy a new house and a new car.

16. A) He had much trouble with his pronunciation.

     B) He began studying English too early.

     C) No one can understand him.

     D) He knew nothing about English.

17. A) Frustration.   B) Joy.

     C) Excitement.   D) Sorrow.

18. A) He likes to go out of town.   B) He can't attend.

     C) He never attends novel reading.   D) He isn't going out of town next week.

Questions 19 to 21 are based on the conversation you have just heard.

19. A) Husband and wife.   B) Doctor and patient.

     C) Teacher and student.   D) Mother and son.

20. A) The woman's parents were coming.

　　B) The man's parents who smoked were coming.

　　C) The man's parents were coming to stop the woman smoking.

　　D) The woman asked her parents to come to stop the man smoking.

21. A) The man didn't like his parents.

　　B) The woman hated the man's parents.

　　C) The man hated smoking.

　　D) The woman hated the man's parents smoking.

Questions 22 to 25 are based on the conversation you have just heard.

22. A) In the office.　　　　　　　　　　B) A shop selling computers.

　　C) A company repairing computers.　　D) On the phone.

23. A) He lost all his data in his work.

　　B) He can't find his data in the computer.

　　C) He can't find the phone number of a company.

　　D) His computer is crashed in the middle of his work.

24. A) She thinks she can fix it for him.

　　B) She knows how to solve problems like that by herself.

　　C) The same problem happened to her before and she solved it.

　　D) She knows what happened to him.

25. A) Take the computer to a small company.

　　B) Call computer company a visit.

　　C) Ask help from the company suggested by the woman.

　　D) Switch off his computer.

Section B

Directions: *In this section, you will hear 3 short passages. At the end of each passage, you will hear some questions. Both the passage and the questions will be spoken only once. After you hear a question, you must choose the best answer from the four choices marked A), B), C) and D). Then mark the corresponding letter on the **Answer Sheet 2** with a single line through the center.*

注意：此部分试题请在**答题卡2**上作答

Passage One

Questions 26 to 28 are based on the passage you have just heard.

26. A) Given by the local government.　　　　B) Born by a large number of bitches.

　　C) Bought from different cities and villages.　　D) Captured over grassland.

27. A) 11-week course for control duty. B) 11-week course for patrol duty.

C) 9-week course for control duty. D) 9-week course for patrol duty.

28. A) Catching run-away criminals. B) Scratching the hidden bombs.

C) Patrolling the dangerous town. D) Drug-sniffing and bomb-sniffing.

Passage Two

Questions 29 to 31 are based on the passage you have just heard.

29. A) Gold was discovered.

B) The transcontinental Railroad was completed.

C) The Golden Gate Bridge was constructed.

D) Telegraph communications were established with the East.

30. A) Two million. B) Three million.

C) Five million. D) Six million.

31. A) Nineteen million dollars. B) Thirty-two million dollars.

C) Thirty-seven million dollars. D) Forty-two million dollars.

Passage Three

Questions 32 to 35 are based on the passage you have just heard.

32. A) Computers have become part of our daily lives.

B) Computers have disadvantages as well as disadvantages.

C) People have different attitudes to computers.

D) More and more families will own computers.

33. A) Computers can bring financial problems.

B) Computers can bring unemployment.

C) Computers can be very useful in families.

D) Computerized robots can take over some unpleasant jobs.

34. A) Computers may change the life they have been accustomed to.

B) Spending too much time on computers may spoil people's relationship.

C) Buying computers may cost a lot of money.

D) Computers may take over from human beings altogether.

35. A) Affectionate. B) Disapproving.

C) Approving. D) Neutral.

Section C

Directions: *In this section, you will hear a passage three times. When the passage is read for the first time, you should listen carefully for its general idea. When the passage is read for the second time, you are required to fill in the blanks numbered from 36 to 43 with the exact words you have just heard. For blanks numbered from 44 to 46 you are required to fill in* *the missing information. You can either use the exact words you have just heard or write down the main points in your own words. Finally, when the passage is read for the third time, you should check what you have written.*

注意：此部分试题在**答题卡2**上，请在**答题卡2**上作答

Part IV   Reading Comprehension (Reading in Depth)   (25 minutes)

Section A

Directions： *In this section, there is a passage with ten blanks. You are required to select one word for each blank from a list of choices given in a word bank following the passage. Read the passage through carefully before making your choices. Each choice in the bank is identified by a letter. Please mark the corresponding letter for each item on **Answer Sheet 2** with a single line through the center. **You may not use any of the words in the bank more than once.***

Questions 47 to 56 are based on the following passage.

Where and how can we get water which is safe to drink? As a rule of thumb you will need to have one gallon, per person, per day for drinking ___47___ only. This is drinking ONLY and does not ___48___ water for cooking, bathing, waste handling or for pets. For ___49___ dog, you need another gallon per day and for a cat about a pint a day.

For those who have made NO ___50___ at all at the time a disaster strikes, there are still a few sources. If you think of it in time, the upper tank on each toilet holds several gallons of water which can be dipped out and drunk. Your hot water heater depending upon its ___51___, will have thirty to fifty gallons of water which you can drain off through the tap at the bottom. The ice cubes in your refrigerator are water which is ___52___. Depending upon your home, you may be able to drain back the water from all the pipes in the house to ___53___ more gallons of drinking water.

What to do if you have used up the water from these safe sources? ___54___ in a time of disaster, consider that ANY water not stored or purchased is polluted. Even a crystal clear stream may be ___55___.

If the water you locate is murky or cloudy, first strain the dirt out with several layers of paper towel, ___56___ cloth or coffee filters. Then purify it with one method or another.

注意：此部分试题请在**答题卡2**上作答

A) each	I) dirty
B) clean	J) available
C) include	K) especially
D) throw	L) physically
E) preparations	M) efforts
F) pleasant	N) deadly
G) purposes	O) size
H) obtain	

Section B

Directions: *There are 2 passages in this section. Each passage is followed by some questions or unfinished statements. For each of them there are four choices marked A), B), C) and D). You should decide on the best choice and mark the corresponding letter on the **Answer Sheet 2** with a single line through the centre.*

Passage one

Questions 57 to 61 are based on the following passage.

I'm usually fairly skeptical about any research that concludes that people are either happier or unhappier or more or less certain of themselves than they were 50 years ago. While any of these statements might be true, they are practically impossible to prove scientifically. Still, I was struck by a report which concluded that today's children are significantly more anxious than children in the 1950s. In fact, the analysis showed, normal children aged 9 to 17 exhibited a higher level of anxiety today than children who were treated for mental illness 50 years ago.

Why are America's kids so stressed? The report cites two main causes: increasing physical isolation—brought on by high divorce rates and less involvement in community, among other things —and a growing perception that the world is a more dangerous place.

Given that we can't turn the clock back, we can still do plenty to help the next generation cope.

At the top of the list is *nurturing* (培育) a better appreciation of the limits of individualism. No child is an island. Strengthening social ties helps build communities and protects individuals against stress.

To help kids build stronger connections with others, you can pull the plug on TVs and computers. Your family will thank you later. They will have more time for face-to-face relationships, and they will get more sleep.

Limit the amount of *virtual* (虚拟的) violence your children are exposed to. It's not just video games and movies; children see a lot of murder and crime on the local news.

Keep your expectations for your children reasonable. Many highly successful people never attended Harvard or Yale.

Make exercise part of your daily routine. It will help you cope with your own anxieties and provide a good model for your kids. Sometimes anxiety is unavoidable. But it doesn't have to ruin your life.

注意：此部分试题请在**答题卡 2** 上作答

57. The author thinks that the conclusions of any research about people's state of mind are_____.

    A) reasonable   B) confusing
    C) illogical   D) questionable

58. What does the author mean when he says, we can't turn the clock back?

    A) It's impossible to slow down the pace of change.

    B) The social reality children are facing can not be changed.

    C) Lessons learned from the past should not be forgotten.

    D) It's impossible to forget the past.

59. According to an analysis, compared with normal children today, children treated as mentally ill 50 years ago_____.

    A) were less isolated physically   B) were probably less self-centered
    C) probably suffered less from anxiety D) were considered less individualistic

60. The first and most important thing parents should do to help their children is_____.

    A) to provide them with a safer environment

    B) to lower their expectations for them.

    C) to get them more involved socially.

    D) to set a good model for them to follow.

61. What conclusion can be drawn from the passage?

    A) Anxiety, though unavoidable, can be coped with.

    B) Children's anxiety has been enormously exaggerated.

    C) Children's anxiety can be eliminated with more parental care.

    D) Anxiety, if properly controlled, may help children become mature.

Passage Two

Questions 62 to 66 are based on the following passage.

Whether the eyes are "the windows of the soul" is debatable; that they are intensely important in interpersonal communication is a fact. During the first two months of a baby's life, the stimulus that produces a smile is a pair of eyes. The eyes need not be real: a mask with two dots will produce a smile. Significantly, a real human face with eyes covered will not motivate a smile, nor will the sight of only one eye when the face is presented in profile. This attraction to eyes as opposed to the nose or month continues as the baby matures. In one study, when American four-year-olds were asked to draw people, 75 percent of them drew people with mouths, but 99 percent of them drew people with eyes.

In Japan, however, where babies are carried on their mother's back, infants do not acquire as much attachment to eyes as they do in other countries. As a result, Japanese adults make little use of the face either to *encode* (把……编码) or *decode* (解码) meaning. In fact, Argyle reveals that the "proper place to focus one's gaze during a conversation in Japan is on the neck of one's conversation partner".

The role of eye contact in a conversational exchange between two Americans is well defined: speakers make contact with the eyes of their listener for about one second, then glance away as they talk, in a few moments they re-establish eye contact with the listener to reassure themselves that their audience is still attentive, then shift their gaze away once more. Listener, meanwhile, keep their eyes on the face of the speaker, allowing themselves to glance away only briefly. It is important they be looking at the speaker at the precise moment when the speaker re-establishes eye contact: if they are not looking, the speaker assumes that they are disinterested and either will pause until eye contact is resumed or will terminate the conversation. Just how critical this eye maneuvering is to the maintenance of conversational flow becomes evident when two speakers are wearing dark glasses: there may be a sort of traffic jam of words caused by interruption, false starts, and unpredictable pauses.

注意：此部分试题请在答题卡 2 上作答.

62. The author is convinced that the eyes are_____.

　　A) of extreme importance in expressing feelings and exchanging ideas

　　B) something through which one can see a person's inner world

　　C) of considerable significance in making conversations interesting

　　D) something the value of which is largely a matter of long debate

63. Babies will not be stimulated to smile by a person_____.

　　A) whose front view is fully perceived

　　B) whose face is seen from the side

　　C) whose face is covered with a mask

　　D) whose face is free of any covering

64. According to the passage, the Japanese fix their gaze on their conversation partner's neck because_____.

　　A) they don't like to keep their eyes on the face of the speaker

　　B) they need not communicate through eye contact

　　C) they don't think it polite to have eye contact

　　D) they didn't have much opportunity to communicate through eye contact in babyhood

65. According to the passage, a conversation between two Americans may break down due to_____.

　　A) one temporarily glancing away from the other

　　B) eye contact of more than one second

　　C) improperly-timed ceasing of eye contact

　　D) constant adjustment of eye contact

66. To keep a conversation flowing smoothly, it is better for the participants_____.
   A) not to wear dark spectacles
   B) not to make any interruptions
   C) not to glance away from each other
   D) not to make unpredictable pauses

Part V   Cloze   (15 minutes)

Directions: *There are 20 blanks in the following passage. For each blank there are four choices marked A), B), C), and D) on the right side of the paper. You should choose the ONE that best fits into the passage. Then mark the corresponding letter on **Answer Sheet 2** with a single line through the center.*

注意：此部分试题请在**答题卡 2** 上作答

In every cultivated language there are two great classes of words which, taken together, comprise the whole vocabulary. First, there are those words __67__ which we become acquainted

in daily conversation, which we __68__, that is to

say, from the __69__ of our own family and from

our familiar associates, and __70__ we should

know and use __71__ we could not read or write.

They __72__ the common things of life, and are the

stock in trade of all who __73__ the language. Such words may be called "popular", since they belong to the people __74__ and are not the exclusive

__75__ of a limited class.

On the other hand, our language __76__ a multitude

of words which are comparatively __77__ used in ordinary conversation. Their meanings are known to every educated person, but there is little __78__ to use them at home or in the market-place.

    Our __79__ acquaintance with them comes

67. A. at	B. with
C. by	D. through
68. A. study	B. imitate
C. stimulate	D. learn
69. A. mates	B. relatives
C. members	D. fellows
70. A. which	B. that
C. those	D. ones
71. A. even	B. despite
C. even if	D. in spite of
72. A. mind	B. concern
C. care	D. involve
73. A. hire	B. apply
C. adopt	D. use
74. A. in public	B. at most
C. at large	D. at best
75. A. right	B. privilege
C. share	D. possession
76. A. consists	B. comprises
C. constitutes	D. composes
77. A. seldom	B. much
C. never	D. often
78. A. prospect	B. way
C. reason	D. necessity
79. A. primary	B. first
C. principal	D. prior

not from our mother's __80__ or from the talk of

our school-mates, __81__ from books that we read,

lectures that we __82__, or the more __83__
conversation of highly educated speakers who are

discussing some particular __84__ in a style

appropriately elevated above the habitual __85__

of everyday life. Such words are called "learned",
and the __86__ between them and the "popular"
words is of great importance to a right understanding
of linguistic process.

80. A. tips   B. mouth
    C. lips   D. tongue
81. A. besides   B. and
    C. or   D. but
82. A. hear of   B. attend
    C. hear from   D. listen
83. A. former   B. formula
    C. formal   D. formative
84. A. theme   B. topic
    C. idea   D. point
85. A. border   B. link
    C. degree   D. extent
86. A. diversion   B. distinction
    C. diversity   D. similarity

Part VI   Translation   (5 minutes)

Directions: *Complete the sentences on **Answer Sheet 2** by translating the Chinese given in brackets into English.*

注意：此部分试题在**答题卡 2** 上；请在**答题卡 2** 上作答

答题卡 1 （Answer Sheet 1）

Part I   Writing   (30 minutes)

Directions: *For this part, you are allowed thirty minutes to write a composition on the topic **Why is Imported Fast-Food So Popular?** You should write no less than 120 words and you should base your composition on the outline (given in Chinese) below:*

1. 快餐在中国非常畅销，例如麦当劳、肯德基等;
2. 洋快餐在中国畅销的原因;
3. 一些人对此现象的片面解释以及自己的看法。

Why is Imported Fast-Food So Popular?

_____
_____
_____
_____
_____

答题卡 1 （Answer Sheet 1）

_____

_____

_____

_____

_____

_____

_____

Part Ⅱ   Reading Comprehension (Skimming and Scanning)   (15 minutes)

1. [Y] [N] [NG]   2. [Y] [N] [NG]   3. [Y] [N] [NG]   4. [Y] [N] [NG]

5. [Y] [N] [NG]   6. [Y] [N] [NG]   7. [Y] [N] [NG]

必须使用黑色签字笔书写，在答题区域内作答，超出以下矩形边框限定区域的答案无效。

8. Ancient Greek culture emphasized _____.

9. Streets and buildings are numbered sequentially in _____.

10. French architect designed a modern Indian city, named _____.

答题卡 2（Answer Sheet 2）

Part III   Section A   Section B

11. [A] [B] [C] [D]   16. [A] [B] [C] [D]   21. [A] [B] [C] [D]   26. [A] [B] [C] [D]   31. [A] [B] [C] [D]
12. [A] [B] [C] [D]   17. [A] [B] [C] [D]   22. [A] [B] [C] [D]   27. [A] [B] [C] [D]   32. [A] [B] [C] [D]
13. [A] [B] [C] [D]   18. [A] [B] [C] [D]   23. [A] [B] [C] [D]   28. [A] [B] [C] [D]   33. [A] [B] [C] [D]
14. [A] [B] [C] [D]   19. [A] [B] [C] [D]   24. [A] [B] [C] [D]   29. [A] [B] [C] [D]   34. [A] [B] [C] [D]
15. [A] [B] [C] [D]   20. [A] [B] [C] [D]   25. [A] [B] [C] [D]   30. [A] [B] [C] [D]   35. [A] [B] [C] [D]

Part III   Section C

必须使用黑色签字笔书写，在答题区域内作答，超出以下矩形边框限定区域的答案无效。

Advertising can be thought of as the (36) _____ of making something known in order to buy or sell goods or (37) _____. Advertising (38) _____ to increase people's awareness and (39). _____ interest. It tries to inform and to (40) _____ . The (41) _____ are all used to spread the message. The (42) _____ offers a fairly cheap method. The cinema and (43) _____ radio are useful for local markets. Television, although more expensive, can be very effective. There can be no doubt (44) _____

_____ .

We might ask (45) _____ ,

since advertising forms part of the cost of production, which has to be covered by the selling price.(46) _____ .

答题卡 2（Answer sheet 2）

Part IV　　　**Section A**　　　　　　**Section B**　　　　**Part V**

47. [A][B][C][D][E][F][G][H][I][J][K][L][M][N][O]　57. [A][B][C][D]　67. [A][B][C][D]　77. [A][B][C][D]
48. [A][B][C][D][E][F][G][H][I][J][K][L][M][N][O]　58. [A][B][C][D]　68. [A][B][C][D]　78. [A][B][C][D]
49. [A][B][C][D][E][F][G][H][I][J][K][L][M][N][O]　59. [A][B][C][D]　69. [A][B][C][D]　79. [A][B][C][D]
50. [A][B][C][D][E][F][G][H][I][J][K][L][M][N][O]　60. [A][B][C][D]　70. [A][B][C][D]　80. [A][B][C][D]
51. [A][B][C][D][E][F][G][H][I][J][K][L][M][N][O]　61. [A][B][C][D]　71. [A][B][C][D]　81. [A][B][C][D]
52. [A][B][C][D][E][F][G][H][I][J][K][L][M][N][O]　62. [A][B][C][D]　72. [A][B][C][D]　82. [A][B][C][D]
53. [A][B][C][D][E][F][G][H][I][J][K][L][M][N][O]　63. [A][B][C][D]　73. [A][B][C][D]　83. [A][B][C][D]
54. [A][B][C][D][E][F][G][H][I][J][K][L][M][N][O]　64. [A][B][C][D]　74. [A][B][C][D]　84. [A][B][C][D]
55. [A][B][C][D][E][F][G][H][I][J][K][L][M][N][O]　65. [A][B][C][D]　75. [A][B][C][D]　85. [A][B][C][D]
56. [A][B][C][D][E][F][G][H][I][J][K][L][M][N][O]　66. [A][B][C][D]　76. [A][B][C][D]　86. [A][B][C][D]

Part VI　　　　　　　**Translation**　　　　　　（5 minutes）

必须使用黑色签字笔书写，在答题区域内作答，超出以下矩形边框限定区域的答案无效。

87. Tobacco is taxed in most countries, _____（除了酒之外）.

88. He enjoyed the meal and _____（而且把它全吃光了）.

89. He _____（把成功归因于）hard work and a bit of luck.

90. _____（首先）, I'd like to welcome you to the meeting.

91. _____（就我所知）, this is the first time a British rider has won the competition.

Model Test Six

Part I   **Writing**   **(30 minutes)**

注意：此部分试题在**答题卡**1上

Part Ⅱ   Reading Comprehension (Skimming and Scanning)   (15 minutes)

*Directions: In this part, you will have 15 minutes to go over the passage quickly and answer the questions on **Answer Sheet 1**.*

For questions 1~7, mark

Y *(for **YES**)*   *if the statement agrees with the information given in the passage;*
N *(for **NO**)*   *if the statement contradicts the information given in the passage;*
NG *(for **NOT GIVEN**)*   *if the information is not given in the passage.*

For questions 8~10, complete the sentence with the information given in the passage.

Wisdom Teeth

Wisdom teeth are the last teeth to erupt. This occurs usually between the ages of 17 and 25. There remains a great deal of controversy regarding whether or not these teeth need to be removed. It is generally suggested that teeth that remain completely buried or un-erupted in a normal position are unlikely to cause harm. However, a tooth may become impacted due to lack of space and its eruption is therefore prevented by *gum* (齿龈), bone, another tooth or all three. If these impacted teeth are in an abnormal position, their potential for harm should be assessed.

What are the Indications for Removing Wisdom Teeth?

Wisdom teeth generally cause problems when they erupt partially through the gum. The most common reasons for removing them are:

Decay

Saliva, bacteria and food particles can collect around and impacted wisdom tooth, causing it, or the next tooth to decay. It is very difficult to remove such decay. Pain and infection will usually follow.

Gum Infection

When a wisdom tooth is partially erupted, food and bacteria collect under the gum causing a local infection. This may result in bad breath, pain, and swelling. The infection can spread to involve the cheek and neck. Once the initial episode occurs, each subsequent attack becomes more frequent and more severe.

Pressure Pain

Pain may also come from the pressure of the erupting wisdom tooth against other teeth. In some cases this pressure may cause the erosion of these teeth.

When is the Best Time to Have Wisdom Teeth Removed?

It is now recommended by specialists that impacted wisdom teeth be removed between the ages of 14 and 22 whether they are causing problems or not. Surgery is technically easier and patients recover much more quickly when they are younger. What is a relatively minor operation at 20 can become quite difficult in patients over 40. Also the risk of complications increases with age and the healing process is slower.

Should a Wisdom Tooth be Removed When an Acute Infection is Present?

Generally, no. Surgery in the presence of infection can cause infection to spread and become more serious. Firstly, the infection must be controlled by local oral *hygiene* (卫生)and *antibiotics* (抗生素).

The Pro's and Con's of Wisdom Tooth Removal Some PRO'S Removing a Wisdom Tooth

Wisdom teeth may be hard to access with your toothbrush. Over time, the accumulation of bacteria, sugars and acids may cause a cavity to form in the tooth. If it is restored with a filling, the *cavity* (洞)may spread and destroy more tooth structure and cause severe consequences to the tooth and surrounding supportive structure.

Due to the difficulty of keeping these teeth clean with your daily home care, bacteria and food debris remaining on the wisdom teeth may present a foul smell-causing bad breath.

A wisdom tooth that is still under the gums in a horizontal position (rather than a vertical position) may exert pressure to the surrounding teeth, causing crowding and crooked teeth. This also may occur if there is not enough space in the mouth for the wisdom tooth. This may warrant braces to repair the damage.

Some Cons of Removing the Wisdom Teeth

Depending on the size shape and position of the tooth, removal can vary from a simple extraction to a more complex extraction. With a simple extraction, there is usually little swelling and/or bleeding. More complex extraction will require special treatment which may result in more bruising, swelling and bleeding. However, your dental professional will provide you with post treatment instructions to minimize these side effects.

Following an extraction, a condition called "dry socket" may occur. If the blood *clot* (凝结) that formed in the extraction area becomes removed, it exposes the underlying bone. This condition is very painful, but resolves after a few days. It is preventable by following the post treatment instructions provided by your dental professional.

The longer you wait and the older you get, there is the potential for more problems to occur. This is because as you get older, the bone surrounding the tooth becomes denser, making the tooth more difficult to remove. The healing process may also be slower.

Post Operative Care:

Do Not Disturb the Wound

In doing so you may invite irritation, infection and/or bleeding. Chew on the opposite side for the first 24 hours.

Do Not Smoke for 12 Hours

Smoking will promote bleeding and interfere with healing.

Do Not Spit or Suck Through a Straw

This will promote bleeding and may remove the blood clot, which could result in a dry socket.

Control of Bleeding

If the area is not closed with stitches, a pressure pack made of folded *sterile* (消毒的) *gauze* (纱布) pads will be placed over the socket. It is important that this pack stay in place to control bleeding and to encourage clot formation. The gauze is usually kept in place for 30 minutes. If the bleeding has not stopped once the original pack is removed, place a new gauze pad over the extraction site.

Control of Swelling

After surgery, some swelling is to be expected. This can be controlled by using the cold packs, which slow the circulation. A cold pack is usually placed at the site of swelling during the first 24 hours in a cycle of 20 minutes on and 20 minutes off. After the first 24 hours, it is advisable to *rinse* (漱口) with warm saltwater every two hours to promote healing. (one teaspoon of salt to eight ounces of warm water).

Medication for Pain Control

Pills such as Aspirin can be used to control minor discomfort following oral surgery. The dentist may prescribe stronger medicines if the patient is in extreme discomfort.

Diet and Nutrition

A soft diet may be prescribed for the patient for a few days following surgery.

Following the removal of your wisdom teeth it is important that you call your dentist if any unusual bleeding, swelling or pain occurs. The first 6~8 hours after the extraction are typically the worst, but are manageable with ice packs and non-prescription pain medication. You should also plan to see your dentist approximately one week later to ensure everything is healing well.

It is very important to talk to your dentist about extraction procedure, risks, possible complications and outcomes of the removal of these teeth. The actual extraction may be done by a dentist or it may be referred to an oral surgeon, who is a specialist. This decision is based on the dentist's preference and the unique features of each individual case.

If you are unsure about whether or not to proceed with the treatment suggested by your dental professional, it is a good idea to get a second opinion. If you decide after consulting with a dentist not to have any teeth extracted, they should be monitored at every dental visit.

注意：此部分试题请在答题卡 1 上作答

1. Dentists have reached an agreement that wisdom teeth should be removed in case it leads to other problems.

2. The best time to remove the impacted wisdom teeth is between the ages of 14 and 22 because surgery is technically easier and patients recover much more quickly.

3. If there is an infection, the first thing to do is to stop it before getting the wisdom tooth removed.

4. Wisdom tooth may become the cause of bad breath.

5. Removing the wisdom teeth involve very complex extraction.

6. After your wisdom tooth is removed, you'd better not spit or suck through a straw.

7. After your wisdom tooth is removed, you should rinse with warm water every morning for two weeks.

8～10 题请见答题卡 1

Part Ⅲ   Listening Comprehension   (35 minutes)

Section A

Directions: *In this section, you will hear 8 short and 2 long conversations. At the end of each conversation, one or more questions will be asked about what was said. Both the conversation and the question will be spoken only once. After each question there will be a pause. During the pause, you must read the four choices marked A), B), C)and D), and decide which is the best answer. Then mark the corresponding letter on the **Answer Sheet 2** with a single line through the center.*

注意：此部分试题请在答题卡 2 上作答

11. A) 1016.　　　　　　　　　　　B) 1060.

　　C) 508.　　　　　　　　　　　D) 580.

12. A) He is going to the hospital.　　B) He is showing his hand.

　　C) He is letting her go.　　　　D) He is offering help.

13. A) A shop assistant.　　　　　B) A sales clerk.

　　C) A waiter.　　　　　　　　D) A telephone operator.

14. A) Father and daughter.　　　B) Uncle and niece.

　　C) Aunt and nephew.　　　　D) Cousins.

15. A) She wasn't invited.　　　　B) She wasn't ready to come.

　　C) She altered her decision.　　D) She forget the invitation.

16. A) The door needs repairing.　　B) He had lost all his keys.

　　C) He couldn't open the door.　　D) He wanted the woman to help him.

17. A) He's rather happy to hear so.　B) He's disappointed to hear so.

　　C) He's unhappy to hear so.　　D) He's surprised to hear so.

18. A) He thought it was a good car.

　　B) He thought it was too noisy.

　　C) He thought there was wrong with the car.

　　D) He didn't like it.

Questions 19 to 21 are based on the conversation you have just heard.

19. A) Finish his paper.　　　　　B) Go to work.

　　C) Cook dinner for his cousin.　D) Go to the auditorium.

20. A) Help with a term paper.　　B) Go to a concert.

　　C) Take care of his cousin.　　D) Arrange his cousin's interview.

21. A) He forgot to ask about her hobbies.　B) He heard she enjoys music concerts.

　　C) He thinks she has enjoyed traveling.　D) He doesn't know what she likes to do.

Questions 22 to 25 are based on the conversation you have just heard.

22. A) Reading.　　　　　　　　B) Sleeping.

　　C) Doing research.　　　　　D) Planning a trip.

23. A) To discuss his trip to Mexico.

　　B) To bring him a message from Professor Grant.

　　C) To ask for help with an anthropology assignment.

　　D) To see what progress he's made on his paper.

24. A) He can't sleep at night.

　　B) He can't find a quiet place to study.

　　C) He can't narrow down his research topic.

　　D) He can't find enough information for his research paper.

25. A) She has been to Mexico.   B) She assigns long research papers.

    C) She teaches cultural anthropology.   D) She collects ancient relies.

Section B

Directions: *In this section, you will hear 3 short passages. At the end of each passage, you will hear some questions. Both the passage and the questions will be spoken only once. After you hear a question, you must choose the best answer from the four choices marked A), B), C) and D). Then mark the corresponding letter on the **Answer Sheet 2** with a single line through the center.*

注意：此部分试题请在**答题卡 2** 上作答

Passage One

Questions 26 to 28 are based on the passage you have just heard.

26. A) Sending them to the shop for some milk.

    B) Telling them a dog has died.

    C) Your shoe lace is undone.

    D) Eating something delicious food on the table.

27. A) Her father lost a pen.

    B) Her father didn't know where his penny was.

    C) She told the father that he lost a penny.

    D) Her father wanted to buy something.

28. A) She fooled her father.

    B) She wanted her father to pick it up.

    C) Her father was looking for the penny.

    D) All of the above were not true.

Passage Two

Questions 29 to 31 are based on the passage you have just heard.

29. A) It is not really a new one.

    B) It is the new one but doesn't work properly.

    C) It is the new one but it is not nice looking.

    D) It is the new one but my friend doesn't like it.

30. A) The milk went sour quickly.

    B) The refrigerator had an unusual smell.

    C) She doesn't check every corner inside.

    D) She wiped the refrigerator out.

31. A) The shop promised to change another one.

    B) The shop promised to repair it if you charge it.

    C) The shop promised to repair it for free it if it broke down in the first three months.

    D) The shop promised to repair it for free it if it broke down in one month.

Passage Three

Questions 32 to 35 are based on the passage you have just heard.

32. A) Because they have a driving license.
    B) Because they have received special training.
    C) Because the traffic conditions in London are good.
    D) Because the traffic system of the city is not very complex.

33. A) Two to four months.   B) About three weeks.
    C) At least half a year.   D) Two years or more.

34. A) Government officers are hard to please.
    B) The learner has to go through several tough tests.
    C) The learner usually fails several times before he passes it.
    D) The driving test usually lasts two months.

35. A) They don't want their present bosses to know what they're doing.
    B) They want to earn money from both jobs.
    C) They cannot earn money as taxi drivers yet.
    D) They look forward to further promotion.

Section C

Directions: *In this section, you will hear a passage three times. When the passage is read for the first time, you should listen carefully for its general idea. When the passage is read for the second time, you are required to fill in the blanks numbered from 36 to 43 with the exact words you have just heard. For blanks numbered from 44 to 46 you are required to fill in the missing information. You can either use the exact words you have just heard or write down the main points in your own words. Finally, when the passage is read for the third time, you should check what you have written.*

注意：此部分试题在**答题卡 2** 上，请在**答题卡 2** 上作答

Part IV   Reading Comprehension (Reading in Depth)   (25 minutes)

Section A

Directions： *In this section, there is a passage with ten blanks. You are required to select one word for each blank from a list of choices given in a word bank following the passage. Read the passage through carefully before making your choices. Each choice in the bank is identified by a letter. Please mark the corresponding letter for each item on Answer Sheet 2 with a single line through the centre. You may not use any of the words in the bank more than once.*

Most of what we think of as our atmosphere is actually the *troposphere* (对流层), that part of the atmosphere __47__ to earth. This is where most of our __48__ happens, and it is the only part of the atmosphere which has enough oxygen and warmth for humans to __49__. This part of the atmosphere is about ten miles thick at the equator and slightly half that __50__ at the poles.

Above the troposphere is the stratosphere, which you have probably visited if you have traveled on an international jet liner. But there is another kind of "jet" at this altitude. Huge rivers of air called "jet streams" flow through the stratosphere, and the stratosphere contains the ozone layer which filters harmful ultraviolet rays which could otherwise make life on earth __51__. Above the stratosphere is the mesosphere, and above that is the ionosphere, which is important for radio communications as signals can be bounced off the ionosphere to __52__ parts of the world.

Many people think that the atmosphere is mostly oxygen, since that is what we __53__. But in fact oxygen makes up only about 21% of the atmosphere, and carbon dioxide makes up less than 1%. Over three quarters of the atmosphere is nitrogen, which was expelled from inside the planet while it was still very volcanically __54__. We have a lot of nitrogen in our bodies, but we do not get it __55__ from the atmosphere. Instead we get our nitrogen from plants which we __56__.

注意： 此部分试题请在**答题卡 2** 上作答

A) height	I) possible
B) breathe	J) carefully
C) farthest	K) different
D) directly	L) survive
E) extinct	M) eat
F) size	N) closest
G) weather	O) active
H) sense	

Section B

Directions: *There are 2 passages in this section. Each passage is followed by some questions or unfinished statements. For each of them there are four choices marked A), B), C) and D). You should decide on the best choice and mark the corresponding letter on the **Answer Sheet 2** with a single line through the centre.*

Passage one

Questions 57 to 61 are based on the following passage.

If Europeans thought a drought was something that happened only in Africa, they know better now. After four years of below-normal rainfall (in some cases only 10 percent of the annual average), vast areas of France, Spain, Portugal, Belgium, Britain and Ireland are dry and barren. Water is so low in the canals of northern France that waterway traffic is forbidden except on weekends. *Oyster*(牡蛎)growers in Britain report a 30 percent drop in production because of the loss of fresh water in local rivers necessary for oyster breeding. In southeastern England, the rolling green hills of Kent have turned so brown that officials have been weighing plans to pipe in water from Wales. In Portugal, farmers in the southern Alentejo region have held prayer meetings for rain—so far, in vain.

Governments in drought-plagued countries are taking drastic measures. Authorities in hard-hit areas of France have banned washing cars and watering lawns. In Britain, water will soon be metered, like gas and electricity. "The English have always taken water for granted," says Graham Warren, a spokesman of Britain's National Rivers Authority, "Now they're putting a price on it." Even a sudden end to the drought would not end the misery in some areas. It will take several years of unusually heavy winter rain, the experts say, just to bring existing water reserves up to their normal levels.

注意：此部分试题请在答题卡 2 上作答

57. What does the author mean by saying "they know better now"?

　　A) They know more about the causes.

　　B) They have a better understanding of the drought in Africa.

　　C) They have realized that the drought in Europe is the most serious one.

　　D) They have realized that drought hit not only Africa but also Europe.

58. The drought in Europe failed to lead to the problem of_____.

　　A) below-normal rainfall　　　　　　　　B) difficult navigation

　　C) a sharp drop in oyster harvest　　　　D) bone-dry hills

59. The British government intends to_____.

　　A) forbid the car-washing service

　　B) increase the price of the water used

　　C) end the misery caused by the drought

　　D) put a price on water

60. Which of the following statements is TRUE according to the passage?

　　A) Germany is the only country free from the drought.

　　B) Water reserves are at their lowest level in years due to the drought.

　　C) The drought is more serious in Britain than in France.

　　D) Europe will not have heavy rain until several years later.

61. Which of the following is the most appropriate title for the passage?

　　A) Europe in Misery　　　　　　　　　　B) Drought Attacks Europe

　　C) Be Economical with Water　　　　　　D) Europe, a Would-be Africa

Passage Two

Questions 62 to 66 are based on the following passage.

Real policeman hardly recognize any resemblance between their lives and what they see on TV—if they ever get home in time. There are similarities, of course, but the cops don't think much of them.

The first difference is that a policeman's real life revolves round the law. Most of his training is in criminal law. He has to know exactly what actions are crimes and what evidence can be used to prove them in court. He has to know nearly as much law as a professional lawyer, and what is more, he has to apply it on his feet, in the dark and rain, running down an alley after someone he wants to talk to.

Little of his time is spent in chatting to charming ladies or in dramatic confrontations with desperate criminals. He will spend most of his working life typing millions of words on thousands of forms about hundreds of sad, unimportant people who are guilty—or not—of stupid, petty crimes.

Most television crime drama is about finding the criminal: as soon as he's arrested, the story is over. In real life, finding criminals is seldom much of a problem, except in very serious cases like murders and terrorist attacks, little effort is spent on searching.

Having made an arrest, a detective really starts to work. He has to prove his case in court and to do that he often has to gather a lot of different evidence. So, as well as being overworked, a detective has to be out at all hours of the day and night interviewing his witnesses and persuading them, usually against their own best interests, to help him.

注意: 此部分试题请在**答题卡 2** 上作答

62. It is essential for a policeman to be trained in criminal law_____.

    A) so that he can catch criminal everywhere

    B) because many of the criminals he has to catch are dangerous

    C) so that he can justify his arrests in court

    D) because he has to know nearly as much about law as a professional lawyer

63. The everyday life of a policeman or detective is_____.

    A) exciting and glamorous   B) dangerous and venturous

    C) devoted mostly to routine matters   D) wasted on unimportant matters

64. When murders and terrorists attacks occur, the police_____.

    A) prefer to wait for the criminal to give himself away

    B) spend a lot of effort on trying to track down the criminals

    C) try to make a quick arrest in order to keep up their reputation

    D) usually fail to produce results

65. Which of the following is TRUE according to the passage?

    A) Generally the detective's work is nothing but to arrest criminals.

    B) Policemen feel that the image of their lives shown on TV is not accurate.

C) People are usually willing to give evidence.

D) Policemen and detectives spend little time at the typewriter.

66. Which of the following is the most suitable title for this passage?

A) Real Life of a Detective

B) Detective's life—Fact and Fantasy

C) The Reality of a Detective

D) Policemen and Detective

Part V   Error Correction   (15 minutes)

Directions: *The passage contains TEN errors. Each indicated line contains a maximum of ONE error. In each case, only ONE word involved: You should proofread the passage and correct it in the following way.*

For a wrong word, underline the wrong word and write the correct one in the blank provided at the end of the line.

For a missing word, mark the position of the missing word with a " ^ " sign and write the word you believe to be missing in the blank provided at the end of the line.

For an unnecessary word, cross the unnecessary word with a slash " / " and put the word in the blank provided at the end of the line.

EXAMPLE

When ^art museum wants a new exhibit,
it never buys things in finished form and hangs
them on the wall. When a natural history museum
wants an exhibition, it must often build it.

(1) __an__

(2) __never__

(3) __exhibit__

注意：此部分试题在**答题卡 2** 上，请在**答题卡 2** 上作答

Part VI   Translation   (5 minutes)

Directions: *Complete the sentences on **Answer Sheet 2** by translating the Chinese given in brackets into English.*

注意：此部分试题在**答题卡 2** 上，请在**答题卡 2** 上作答

答题卡 1 （Answer Sheet 1）

Part I　　　　　　　　　　　Writing　　　　　　　　　(30 minutes)

Directions: *For this part, you are allowed thirty minutes to write a composition on the topic **Internet in China**. You should write no less than 120 words and you should base your composition on the outline (given in Chinese) below:*

1. 随着知识经济时代的到来，因特网已走进我们的生活；
2. 因特网能够在中国普及的原因和它的用途；
3. 由于因特网的优势，它将在中国越来越受欢迎。

Internet in China

_____

_____

_____

_____

_____

_____

_____

答题卡 1 （Answer Sheet 1）

_____

_____

_____

_____

_____

_____

_____

_____

_____

_____

_____

_____

_____

_____

Part Ⅱ   Reading Comprehension (Skimming and Scanning)   (15 minutes)

1. [Y] [N] [NG]   2. [Y] [N] [NG]   3. [Y] [N] [NG]   4. [Y] [N] [NG]

5. [Y] [N] [NG]   6. [Y] [N] [NG]   7. [Y] [N] [NG]

必须使用黑色签字笔书写，在答题区域内作答，超出以下矩形边框限定区域的答案无效。

8. If the position of the impacted teeth is abnormal, it is advisable to weigh_____.

9. Even if you have decided not to get your wisdom teeth removed, you still have to have them_____.

10. The painful time during the first 6～8 hours after the removal of the tooth can be controlled with _____.

答题卡 2（Answer Sheet 2）

Part III　　　　**Section A**　　　　　　　　　　　**Section B**

11. [A] [B] [C] [D]　16. [A] [B] [C] [D]　21. [A] [B] [C] [D]　26. [A] [B] [C] [D]　31. [A] [B] [C] [D]

12. [A] [B] [C] [D]　17. [A] [B] [C] [D]　22. [A] [B] [C] [D]　27. [A] [B] [C] [D]　32. [A] [B] [C] [D]

13. [A] [B] [C] [D]　18. [A] [B] [C] [D]　23. [A] [B] [C] [D]　28. [A] [B] [C] [D]　33. [A] [B] [C] [D]

14. [A] [B] [C] [D]　19. [A] [B] [C] [D]　24. [A] [B] [C] [D]　29. [A] [B] [C] [D]　34. [A] [B] [C] [D]

15. [A] [B] [C] [D]　20. [A] [B] [C] [D]　25. [A] [B] [C] [D]　30. [A] [B] [C] [D]　35. [A] [B] [C] [D]

Part III　　　　　　　　　　**Section C**

必须使用黑色签字笔书写，在答题区域内作答，超出以下矩形边框限定区域的答案无效。

　　In the north of Scotland, there is a deep, dark lake (36)_____ by mountains. This is Loch Ness. Loch is the (37) _____ word for "lake". Loch Ness is related to a big and strange creature that was said to live in the lake. (38) _____no one ever got a good look at it, the local people believed in this creature. They (39) _____it must be some kind of fish, since it lived in the lake.

　　Before the 1930s, few outsiders had heard of the creature. Then a road was built along Loch Ness and many (40) _____ came to see for themselves. Some believed they had caught sight of it. Many newspapers (41) _____stories about the creature.

　　These stories made the (42) _____ famous. But some readers (43) _____ it a joke. To them, this strange thing was an invented animal, something they might see in a movie. (44)_____.

Many people thought they had seen part of it. The parts added up to a very strange creature indeed. It was said to be 20 or 30 or 50 feet long. (45)_____

_____.

There was a long neck with a small head. Sometimes the back looked like a boat turned upside down. At other times it had one, two or three humps like a camel. (46) _____

_____.It never

attacked people, and any noise caused it to disappear.

答题卡 2（Answer Sheet 2）

Part Ⅳ

Section A

47.[A][B][C][D][E][F][G][H][I][J][K][L][M][N][O]
48.[A][B][C][D][E][F][G][H][I][J][K][L][M][N][O]
49.[A][B][C][D][E][F][G][H][I][J][K][L][M][N][O]
50.[A][B][C][D][E][F][G][H][I][J][K][L][M][N][O]
51.[A][B][C][D][E][F][G][H][I][J][K][L][M][N][O]
52.[A][B][C][D][E][F][G][H][I][J][K][L][M][N][O]
53.[A][B][C][D][E][F][G][H][I][J][K][L][M][N][O]
54.[A][B][C][D][E][F][G][H][I][J][K][L][M][N][O]
55.[A][B][C][D][E][F][G][H][I][J][K][L][M][N][O]
56.[A][B][C][D][E][F][G][H][I][J][K][L][M][N][O]

Section B

57.[A][B][C][D]
58.[A][B][C][D]
59.[A][B][C][D]
60.[A][B][C][D]
61.[A][B][C][D]
62.[A][B][C][D]
63.[A][B][C][D]
64.[A][B][C][D]
65.[A][B][C][D]
66.[A][B][C][D]

Part V   Error Correction   (15 minutes)

One of the best ways to celebrate Mother's Day is to give
your mom the day off. Let her take it ease and relax while the rest   67_____
of the family does the work.

Many families began Mother's Day with breakfast in bed.   68_____
Usually Dad and the kids will let mom sleep later, they go into   69_____
the kitchen and prepare her favorite meal. A Mother's Day breakfast
can consist anything your mom likes. After the food is cooked,   70_____
they arrange everything nice on a tray. Don't forget the vase with a   71_____
single flower. With spring here, the children can pick a *tulip* (郁金香)
or *daffodil* (水仙花) from the garden outside.

When everything are ready they carefully carry the tray and   72_____
mom's favorite sections from the kitchen to her bedroom. Cards and
small present from the children can be placed on the tray before it is   73_____
present to mom in bed.   74_____

Many families make a special Mother's Day dinner at home   75_____
and take mom out to her favorite restaurant for a meal. It is a good   76_____
day to let your mom relax and let her see how a wonderful family
she has.

必须使用黑色签字笔书写，在答题区域内作答，超出以下矩形边框限定区域的答案无效。

77. _____ (平均来看), men smoke more cigarettes than women.

78. Your losses in trade this year are nothing_____ (与我的相比).

79. Not only _____ (他向我收费过高)，but he didn't do a good repair job either.

80. Mary has to work with people of all ages _____ (各行各业).

81. The children _____ (退学) and went to work.

Model Test Seven

Part I   Writing   (30 minutes)

注意：此部分试题在**答题卡 1** 上

Part II   Reading Comprehension(Skimming and Scanning)   (15 minutes)

Directions: *In this part, you will have 15 minutes to go over the passage quickly and answer the questions on **Answer Sheet 1**.*

For questions 1~7, mark

    Y *(for YES)*   *if the statement agrees with the information given in the passage,*

    N *(for NO)*   *if the statement contradicts the information given in the passage;*

    NG *(for NOT GIVEN)*   *if the information is not given in the passage.*

For questions 8~10, complete the sentences with the information given in the passage.

What causes earthquakes? The earth is formed of layers. The surface of the earth, about 100 kilometers thick, is made of large pieces. When they move against each other, an earthquake happens. A large movement causes a violent earthquake, but a small movement causes a mild one.

Earthquakes last only a few seconds. The rolling movements are called seismic waves. The seismic waves start in one place, called the epicenter, and roll outward. A seismic wave travels around the earth in about twenty minutes. Usually, an earthquake is strong enough to cause damage only near its epicenter.

However, epicenters at the bottom of the ocean create huge sea waves as tall as 15 meters. These waves cross the ocean in several hours. Rushing toward land, they destroy small islands and ships path. When they hit land, they flood coastal areas far from the epicenter of the earthquake. In 1868, a wave in Japan killed 27,000 people.

After an earthquake happens, people can die from lack of food, water, and medical supplies. The amount of destruction caused by an earthquake depends on where it happens, what time it happens, and how strong it is. It also depends on types of buildings, soil conditions, and population. Of the 6,000 earthquakes in the world each year, only about fifteen cause great damage and many deaths.

In 1556, an earthquake in northern China killed 830,000 people—the most in history. There was no way to measure its strength. In 1935, scientist started using the Richter Scale to measure seismic waves. A seriously destructive earthquake measures 6.5 or higher on the Richter Scale.

How can scientists predict earthquakes? Earthquakes are not just scattered anywhere but happen in certain areas, places where pieces of the Earth's surface meet. This pattern causes them to shake the same places many times. For example, earthquakes often occur on the west coast of North and South American, around the Mediterranean Sea, and along the Pacific coast of Asia.

Another way to predict earthquakes is to look for changes in the earth's surface, like a sudden drop of water level in the ground. Some people say animals can predict earthquakes. Before earthquakes, people have seen chickens sitting in trees, fish jumping out of the water, snakes leaving their holes, and other animals acting strangely.

On February 4, 1975, scientists predicted an earthquake in northeastern China and told people in the earthquake zone to leave the cities. More than a million people moved into the surrounding countryside, into safe, open fields away from buildings. That afternoon, the ground rolled and shook beneath the people's feet. In seconds, 90 percent of the buildings in the city of Haicheng were destroyed. The decision to tell the people to leave the cities saved 10,000 lives.

However, more than a year later, on July 28, 1976, the scientists were not so lucky. East of Beijing, Chinese scientists were discussing a possible earthquake. During their meeting, the worst earthquake in modern times hit. Estimates of deaths ranged from 250,000 to 695,000. The earthquake measured 7.9 on the Richter Scale.

Earthquakes often come together with volcanic eruptions. In late 1984, strong earthquakes began shaking the Nevado del Ruiz volcano in Colombia every day. On November 14, 1985, it erupted. A nearby river became a sea of mud that buried four towns. This disaster killed more than 2,100 people.

Mexico City has frequent earthquakes. An earthquake there on September 19, 1985, measured 8.1 on the Richter Scale and killed 7,000 people. Most victims died when buildings fell on them.

San Francisco, California, also has frequent earthquakes. However, newer buildings there are built to be safe in earthquakes. Therefore, when an earthquake measuring 7.1 on the Richter Scale hits northern California on October 17, 1989, only 67 people were killed. The earthquake hit in the afternoon, when thousands of people were driving home from work. Freeways and bridges broke and fell. Buried under the lavers of the Oakland Freeway, people were crushed in their flattened cars. Explosions sounded like thunder as older buildings seemed to burst apart along with the freeways. As the electric power lines broke from the falling bridges and buildings, the sky, covered with huge clouds of black dust, appeared to be filled with lightning. Water rushed into the streets from broken pipes and mixed with gas from broken gas lines, causing more explosions.

Emergency workers had to cope with medical problems. Everyone worked together to save survivors and comfort victims. The next day, the disaster sites looked terrible. Victims couldn't find their houses, their cars, or even their streets. Boats were destroyed, and debris covered the surface of the sea. There was no water, no electricity, no telephone only the smell of garbage floating in melted ice in refrigerators open to the sun. Losses and property damage from the earthquake amounted to millions of dollars.

Seismology is the study of earthquakes, and a seismologist is a scientist who observes earthquakes. Seismologists have given us valuable knowledge about earthquakes. Their equipment measures the smallest vibration on the surface of the earth. They are trying to find ways to use knowledge about earthquakes to save lives and to help solve the world's energy shortage. The earth's natural activity underground creates energy in the form of heat. Geothermal means earth heat. This geothermal energy could be useful. However, if we take natural hot water out of the earth in earthquake zones, we might cause earthquakes.

People live in earthquake zones because of natural beauty, productive soil, and large existing centers of population. However, people who live there should expect earthquakes. They should be prepared to protect their lives and property. They must build safer buildings and roads. Hospitals and electric power stations must be built as far as possible from probable earthquake sites. When an earthquake starts, people must run to open ground or stay in protected areas like doorways or even bathtubs.

If seismologists could predict earthquakes, we could save about 20,000 human lives each year. Human can control many things about nature, but we can not control earthquakes.

注意：此部分试题请在答题卡 1 作回答

1. Today scientists know something about the causes of earthquakes.
2. More people are killed by huge sea waves than by buildings falling.
3. The vast majority of the world's earthquakes are mild.
4. An earthquake in 1989 destroyed the city of Oakland.
5. Seismologists can measure the size of sea waves.
6. Removing water from underground may cause earthquakes.
7. The passage gives a general description of the earthquakes' destruction.

8～10 题请见答题卡 1

Part III   Listening Comprehension   (35 minutes)

Section A

Directions: *In this section, you will hear 8 short and 2 long conversations. At the end of each conversation, one or more questions will be asked about what was said. Both the conversation and the question will be spoken only once. After each question there will be a pause. During the pause, you must read the four choices marked A), B), C)and D), and decide which is the best answer. Then mark the corresponding letter on the **Answer Sheet 2** with a single line through the center.*

11. A) 4.   B) 3.
    C) 7.   D) 8.

12. A) She baked the cake herself.   B) She bought it from the shop.
    C) She wanted the man to bake it for her.   D) The bakery baked the cake.

13. A) To buy some potatoes.   B) To pass him some potatoes.
    C) To have some potatoes.   D) To help him cook some potatoes.

14. A) Jim looks nice in anything.   B) Jim doesn't look nice in the new shirt.
    C) Jim does not look nice in the old shirt.   D) He wants Jim to lend him that shirt.

15. A) The rain has stopped.   B) She wants to soak her clothes.
    C) She is looking for her clothes.   D) It's raining heavily.

16. A) She will give him something to write.

    B) She doesn't know who took away his notebook.

    C) She wants to borrow some paper for him.

    D) He can borrow a notebook from her.

17. A) She wants to go in another day.   B) It will depend on the weather.
    C) It's a nice day to go to class.   D) Some of them can go swimming.

18. A) He is very surprised to have such a bad weather.

    B) He doesn't think the train will be late.

    C) The train is late because of bad weather.

    D) Nobody will be surprised that the train is late.

Questions 19 to 21 are based on the conversation you have just heard.

19. A) Start a new program at State College.   B) Study at a different school.
    C) Work at an art gallery.   D) Move to a warmer state.

20. A) Since the fall.   B) Since the summer.
    C) For a year and a half.   D) For three years.

21. A) Use her professor as references.

    B) Improve her grades.

    C) Think more positively about the State College program.

    D) Write to the head of the art department.

Questions 22 to 25 are based on the conversation you have just heard.

22. A) Teacher and student.   B) Employer and employee.
    C) Friends.   D) Classmates.

23. A) In a hospital.   B) In Jerry's office.
    C) In Ms. Sherwin's office.   D) In Jenny's home.

24. A) To ask for a few days off.   B) To talk about his daughter.
    C) To send his wife to hospital.   D) To talk about his job.
25. A) Tomorrow.   B) The day after tomorrow.
    C) In a few days.   D) Tonight.

Section B

Directions: *In this section, you will hear 3 short passages. At the end of each passage, you will hear some questions. Both the passage and the questions will be spoken only once. After you hear a question, you must choose the best answer from the four choices marked A), B), C) and D). Then mark the corresponding letter on the **Answer Sheet 2** with a single line through the center.*

注意：此部分试题请在答题卡 2 上作答

Passage One

Questions 26 to 28 are based on the passage you have just heard.

26. A) Have to buy a special electronic ticket.

    B) Have to travel a long way to visit the university.

    C) Need an expensive device designed especially for the museum.

    D) Need a computer linked to a telephone.

27. A) Provide a place for computer artists to show their work.

    B) Sell the art works more easily.

    C) Save space of museums for other purposes.

    D) Provide more fun for the artists.

28. A) It helps a computer artist to record his pictures electronically.

    B) It helps a computer artist to send his pictures to others.

    C) It helps a computer artist to print pictures on paper.

    D) It helps a computer artist to connect his computer to the art museum.

Passage Two

Questions 29 to 31 are based on the passage you have just heard.

29. A) 4 years.   B) 5 years.
    C) 8 years.   D) at least 9 years.
30. A) Biology.   B) Chemistry.
    C) Philosophy.   D) Medicine.
31. A) Each student must pass a national examination.

    B) Students who do best in the studies have a greater chance.

    C) They can seek to enter a number of medical schools

    D) Chances to gain the entrance are many.

Passage Three

Questions 32 to 35 are based on the passage you have just heard.

32. A) Guarding the coats of the United States.

    B) Being part of the United States Navy.

    C) Guiding people along the coast.

    D) Protecting people from army attack.

33. A) Enforcing laws controlling navigation, shipping, immigration and fishing.

    B) Enforcing laws affecting the privately-owned boats in the U.S. .

    C) Searching for missing boats and rescuing people.

    D) Training people to good swimmers along the beach.

34. A) 17,000.   B) 1,700.

    C) 70,000.   D) 7,000.

35. A) Dangerous.   B) Hard.

    C) Exciting.   D) Dull.

Section C

Directions: *In this section, you will hear a passage three times. When the passage is read for the first time, you should listen carefully for its general idea. When the passage is read for the second time, you are required to fill in the blanks numbered from 36 to 43 with the exact words you have just heard. For blanks numbered from 44 to 46 you are required to fill in the missing information. You can either use the exact words you have just heard or write down the main points in your own words. Finally, when the passage is read for the third time, you should check what you have written.*

注意：此部分试题在**答题卡 2** 上，请在**答题卡 2** 上作答

Part IV   Reading Comprehension (Reading in Depth)   (25 minutes)

Section A

Directions: *In this section, there is a passage with ten blanks. You are required to select one word for each blank from a list of choices given in a word bank following the passage. Read the passage through carefully before making your choices. Each choice in the bank is identified by a letter. Please mark the corresponding letter for each item on **Answer Sheet 2** with a single line through the centre. **You may not use any of the words in the bank more than once.***

The ability to express oneself and to communicate is extremely important for public speakers. It is also important for those who want to ___47___ themselves well in a society that is full of ___48___ competition. How can we gain such ability, especially for those who are born timid and shy?

Well, first of all, your   49   about the importance of public speaking is absolutely correct. People need to be able to   50   themselves effectively, whether they are going to be a journalist such as you are, whether they are going to be a businessman or businesswoman, whether they are   51   education and to be a teacher. And if people are going to law, they have to   52   effectively all the time, and much of the communication take the form of public speaking. Some take written form, for example, journalism. We have print journalism and broadcast journalism, but the   53   of effective communication are substantially the same. Of course, all of those are forms of communication. So if you can develop the skills of an effective public speaker, you can use those skills not just in the public speaking but in other forms of communication. You can use them in writing, you can use them in conversation. I talked a few days ago with a reporter from XinHua News   54   and he said, " you know, what you are talking about is exactly the same thing that we do in articles or journalism. You have a lead, you get readers'   55 ; you have a main point, you develop the main point; you have the second point, you develop the point; you have conclusion that   56   everything up." So the skills of public speaking can be transferred or related to other areas of communication.

注意：此部分试题请在**答题卡 2** 上作答

A) efficient	I) express
C) constitute	J) principles
C) constitute	K) Agency
D) attention	L) communicate
E) fierce	M) purpose
F) establish	N) point
G) Association	O) wild
H) going in for	

Section B

Directions： *There are 2 passages in this section. Each passage is followed by some questions or unfinished statements. For each of them there are four choices marked A), B), C) and D). You should decide on the best choice and mark the corresponding letter on the **Answer Sheet 2** with a single line through the centre.*

Passage one

Questions 57 to 61 are based on the following passage.

In contemporary Asia most countries have granted equal status to women legally. However, in many countries the social mixing of men and women is still viewed with distrust and a woman is expected to remain in the background. If a woman has a profession, it is almost as if she is an abnormal kind of woman.

Fortunately, social attitudes do not last forever. Although the change is slow, there is some change in Asian opinion about women having jobs. Medicine, nursing and teaching have the longest history. Women doctors were necessary in some countries where male doctors were not permitted, for social or religious reasons, to see women patients. Another recognized profession for women is to work in offices and some have reached very high positions. This tends to suggest that in some areas men's refusal to accept the professional ability—and equality—of women is gradually being eroded.

But being allowed to have a job is not enough. True liberation can't exist until there is wider social equality. This is not a one-way process. Women need to be educated to understand the meaning of their right, particularly in countries where the status of women has been low in the past. Yet in just these countries a widely held belief is that if women are educated they will become less "womanly".

In many ways the fight for the real liberation of women throughout Asia has only just begun. Although women now have political and legal rights, practicing these rights in everyday life is a far more difficult matter. The frequently heard remark "You've got your rights now, what more do you want?" sums up this feeling. Asian women now want to put these rights into practice and it is here that they meet with opposition. "Women," one hears men say, "are getting too forward these days."

注意： 此部分试题请在**答题卡 2** 上作答

57. The so-called "normal" women should_____.

    A) be independent and educated   B) stay at home all the day

    C) stay in the background socially   D) have a stable and good job

58. Medicine, nursing and teaching are jobs_____.

    A) recognized for women   B) for men and women

    C) beyond women's ability   D) for professional women

59. The word "eroded" in Para.2 probably means_____.

    A) strengthened   B) recognized   C) reduced   D) changed

60. What's the author's attitude towards women's liberation?

    A) Indifferent   B) Supportive   C) Disappointed   D) Humorous

61. What's the main idea of this article?

    A) Asian women have gained their rights.

    B) There are many good jobs for women.

    C) Asian women want to put their rights into practice.

    D) The liberation of Asian women.

Passage Two

Questions 62 to 66 are based on the following passage.

With a tremendous roar from its rocket engine, the satellite is sent up into the sky. Minutes later, at an altitude of 300 miles, this tiny electronic moon begins to orbit about the earth. Its radio begins to transmit an astonishing amount of information about the satellite's orbital path, the amount of radiation it detects, and the presence of meteorites. Information of all kind races back to the earth. No human being could possibly copy down all these facts, much less remember and organize them. But an electronic computer can.

The marvel of the machine age, the electronic computer has been in use only since 1946. It can do simple computations—add, subtract, multiply and divide—with lighting speed and perfect accuracy. Some computers can work 500,000 times faster than any person can.

Once it is given a "program"—that is, a carefully worked-out set of instructions devised by a technician trained in computer language—a computer can gather a wide range of information for many purposes. For the scientist it can get information from outer space or from the depth of the ocean. In business and industry the computer prepares factory inventories, keeps track of sales trends and production needs, mails dividend checks, and makes out company payrolls. It can keep bank accounts up to date and make out electric bills. If you are planning a trip by plane, the computer will find out what to take and what space is available.

Not only can the computer gather facts, it can also store them as fast as they are gathered and can pour them out whenever they are needed. The computer is really a high-powered "memory" machine that "had all the answers"—or almost all. Besides gathering and storing information, the computer can also solve complicated problems that once took months for people to do.

At times computers seem almost human. They can "read" hand-printed letters, play chess, compose music, write plays and even design other computers. Is it any wonder that they are sometimes called "thinking" machines?

Even though they are taking over some of the tasks that were once accomplished by our own brains, computers are not replacing us at least not yet. Our brain has more than 10 million cells. A computer has only a few hundred thousand parts. For some time to come, then, we can safely say that our brains are at least 10,000 times more complex than a computer. How we use them is for us, not the computer, to decide.

注意：此部分试题请在**答题卡 2** 上作答

62. In the first paragraph, the author thinks an electronic computer can_____.

   A) copy down all the facts

   B) remember all the facts

C) organize the facts and everything

D) copy down, remember and organize all the facts

63. "Program" means_____.

A) a plan of what is to be done

B) a complete show on a TV station at a fixed time table

C) a scheduled performance

D) series of coded instructions to control the operations of a computer

64. The computer is a high powered "memory" machine, which_____.

A) has all the ready answers—or almost all to any questions

B) can remember everything

C) has all the answers—or almost to all the information that has been stored

D) can store everything and work for you

65. "Thinking" machines suggest that _____.

A) they can "read" hand printed letters etc

B) they can't think, but can do something under human control

C) they even design other computers

D) they really can think and do many other jobs

66. Why can't computers do whatever they want to do?

A) Because some computers can't work 500,000 times faster than any person can.

B) Because they normally have a few hundred thousand parts.

C) Because human brains are at least 10,000 times more complex than any computers.

D) Because how a computer works is decided by human.

Part V   Cloze   (15 minutes)

Directions: *There are 20 blanks in the following passage. For each blank there are four choices marked A), B), C), and D) on the right side of the paper. You should choose the ONE that best fits into the passage. Then mark the corresponding letter on **Answer Sheet 2** with a single line through the center.*

注意：此部分试题请在**答题卡2**作答

One summer night, on my way home from work I decided to see a movie. I knew the theatre would be air-conditioned and I couldn't face my __67__ apartment.

Sitting in the theatre I had to look through the __68__ between the two tall beads in front of me. I had to keep changing the __69__ every time she I leaned over to talk to him. __70__ he learned over to kiss her. Why do Americans display such __71__ in a public place?

I thought the movie would be good for my English, but __72__ it turned out, it was an Italian movie. __73__ about an

hour I decided to give up on the movie and __74__ on my *popcorn*(爆玉米花). I've never understood why they give you so much popcorn! It tasted pretty good, __75__. After a while I

heard __76__ more of the romantic-sounding Italians, I just

heard the __77__ of the popcorn *crunching*（咀嚼）between

my teeth. My thought started to __78__. I remembered when

I was in *South Korea* (韩国), I __79__ to watch Kajak on TV frequently. He spoke perfect Korean. I was really amazed. He seemed like a good friend to me, __80__ I saw him again in

New York speaking __81__ English instead of perfect Korean.

He didn't even have a Korean accent and I __82__ like I had been betrayed.

When our family moved to the United States six years ago, none of us spoke any English. __83__ we had begun to learn a few words, my mother suggested that we all should speak English at home. Everyone agreed, but our house became very __84__ and we all seemed to avoid each other. We sat at the

67.	A. hot	B. warm	
	C. cool	D. heated	
68.	A. crack	B. opening	
	C. break	D. blank	
69.	A. view	B. angle	
	C. space	D. aspect	
70.	A. whenever	B. and	
	C. while	D. or	
71.	A. attention	B. affection	
	C. motion	D. attraction	
72.	A. what	B. when	
	C. as	D. anise	
73.	A. after	B. within	
	C. for	D. over	
74.	A. fix	B. chew	
	C. taste	D. concentrate	
75.	A. too	B. though	
	C. still	D. certainly	
76.	A. no	B. few	
	C. any	D. much	
77.	A. tone	B. voice	
	C. sound	D. rhythm	
78.	A. wonder	B. imagine	
	C. depart	D. wander	
79.	A. used	B. happened	
	C. turned	D. enjoyed	
80.	A. then	B. until	
	C. because	D. therefore.	
81.	A. artificial	B. informal	
	C. practical	D. perfect	
82.	A. looked	B. felt	
	C. appeared	D. seemed	
83.	A. if	B. while	
	C. once	D. before	
84.	A. quiet	B. empty	

dinner table in silence, preferring that to _85_ in a difficult language. Mother tried to say something in English but it _86_ out all wrong and we all burst into laughter and decided to forget it! We're been speaking Korean at home ever since.

C. stiff   D. calm

85. A. saying   B. uttering
    C. telling   D. speaking

86. A. made   B. worked
    C. got   D. came

Part VI   Translation   (5 minutes)

Directions: *Complete the sentences on Answer Sheet 2 by translating the Chinese given in brackets into English.*

注意：此部分试题在**答题卡2**上，请在**答题卡2**上作答

答题卡 1 （Answer Sheet 1）

Part I   Writing   (30 minutes)

Directions: *For this part, you are allowed thirty minutes to write a composition on the topic **Desk Culture**. You should write no less than 120 words and you should base your composition on the outline (given in Chinese) below:*

1. 什么是课桌文化（学生在课桌上留下的文字或符号）；
2. 课桌文化的内容及产生的原因；
3. 你的看法。

Desk Culture

_____

_____

_____

_____

_____

_____

答题卡 1 （Answer Sheet 1）

_____
_____
_____
_____
_____
_____
_____
_____
_____
_____
_____
_____
_____

Part Ⅱ   Reading Comprehension (Skimming and Scanning)   (15 minutes)

1. [Y] [N] [NG]   2. [Y] [N] [NG]   3. [Y] [N] [NG]   4. [Y] [N] [NG]

5. [Y] [N] [NG]   6. [Y] [N] [NG]   7. [Y] [N] [NG]

必须使用黑色签字笔书写，在答题区域内作答，超出以下矩形边框限定区域的答案无效。

8. The death toll of the earthquake in northern China—the most in history—reached_____.

9. Earthquakes often come together with _____.

10. When an earthquake occurs, people must run to open ground or stay in safe areas like_____
_____.

答题卡 2 (Answer Sheet 2)

Part III

Section A

11. [A] [B] [C] [D]   16. [A] [B] [C] [D]   21. [A] [B] [C] [D]
12. [A] [B] [C] [D]   17. [A] [B] [C] [D]   22. [A] [B] [C] [D]
13. [A] [B] [C] [D]   18. [A] [B] [C] [D]   23. [A] [B] [C] [D]
14. [A] [B] [C] [D]   19. [A] [B] [C] [D]   24. [A] [B] [C] [D]
15. [A] [B] [C] [D]   20. [A] [B] [C] [D]   25. [A] [B] [C] [D]

Section B

26. [A] [B] [C] [D]   31. [A] [B] [C] [D]
27. [A] [B] [C] [D]   32. [A] [B] [C] [D]
28. [A] [B] [C] [D]   33. [A] [B] [C] [D]
29. [A] [B] [C] [D]   34. [A] [B] [C] [D]
30. [A] [B] [C] [D]   35. [A] [B] [C] [D]

Part III   Section C

必须使用黑色签字笔书写，在答题区域内作答，超出以下矩形边框限定区域的答案无效。

The (36)_____ was open at 10 a.m. It was 9:30 and already the pilgrims had formed a (37) _____ I asked a lady from Ohio why she had come. "For Shakespeare," she said. "Isn't that why you came?"

"Not (38) _____," I said. "I was born here. I'm visiting my family."

"You were born here?" she said, as if only Shakespeare had the right.

It was my first time in many years. Long ago I had (39) _____ to America. Now I was visiting places in which I had taken little (40) _____ before: the birthplace for example. I had passed it perhaps a hundred times without a thought of going in. Now it would cost me just under two pounds, about $3. An even stranger experience was buying a ticket to the school two (41) _____of my family had attended. Shakespeare had gone there, though 350 years before.

It was a good school, but I was (42) _____ in being sent to a better one. "Better than Shakespeare's?" asked an American to whom I had confided. "I don't see how that could be (43) _____,"he had muttered before turning away.(44)_____ .However, in my current tourist status, that would have to be changed (45)_____.I must learn to refer to him as "the bard", and not as "Will" in the familiar way, and never a "Willie the Shake", which is the inelegant but customary nickname of some of the younger generation. This was no problem. Shakespeare worship had begun before my day. Every building with Shakespeare connections was preserved. (46)_____ .

答题卡 2（Answer Sheet 2）

Part IV	Section A		Section B		Part V

Section A
47. [A][B][C][D][E][F][G][H][I][J][K][L][M][N][O]
48. [A][B][C][D][E][F][G][H][I][J][K][L][M][N][O]
49. [A][B][C][D][E][F][G][H][I][J][K][L][M][N][O]
50. [A][B][C][D][E][F][G][H][I][J][K][L][M][N][O]
51. [A][B][C][D][E][F][G][H][I][J][K][L][M][N][O]
52. [A][B][C][D][E][F][G][H][I][J][K][L][M][N][O]
53. [A][B][C][D][E][F][G][H][I][J][K][L][M][N][O]
54. [A][B][C][D][E][F][G][H][I][J][K][L][M][N][O]
55. [A][B][C][D][E][F][G][H][I][J][K][L][M][N][O]
56. [A][B][C][D][E][F][G][H][I][J][K][L][M][N][O]

Section B
57.[A][B][C][D]
58.[A][B][C][D]
59.[A][B][C][D]
60.[A][B][C][D]
61.[A][B][C][D]
62.[A][B][C][D]
63.[A][B][C][D]
64.[A][B][C][D]
65.[A][B][C][D]
66.[A][B][C][D]

67.[A][B][C][D]
68.[A][B][C][D]
69.[A][B][C][D]
70.[A][B][C][D]
71.[A][B][C][D]
72.[A][B][C][D]
73.[A][B][C][D]
74.[A][B][C][D]
75.[A][B][C][D]
76.[A][B][C][D]

Part V
77. [A][B][C][D]
78. [A][B][C][D]
79. [A][B][C][D]
80. [A][B][C][D]
81. [A][B][C][D]
82. [A][B][C][D]
83. [A][B][C][D]
84. [A][B][C][D]
85. [A][B][C][D]
86. [A][B][C][D]

Part VI   Translation   (5 minutes)

必须使用黑色签字笔书写，在答题区域内作答，超出以下矩形边框限定区域的答案无效。

87. The substance does not dissolve in water_____(不管是否加热).

88. There are several people like me who would _____(以某人为榜样) over anything else.

89. You should _____(抽出时间和力量来帮助别人)when necessary.

90. You are a popular girl, Linda, and many younger ones _____(尊敬).

91. No one could _____ (代替) her father.

Model Test Eight

Part I   **Writing**   **(30 minutes)**

注意：此部分试题在答题卡 1 上

Part II   **Reading Comprehension (Skimming and Scanning)**   **(15 minutes)**

Directions: *In this part, you will have 15 minutes to go over the passage quickly and answer the questions on **Answer Sheet 1**.*

For questions 1~7, mark

     Y *(for **YES**)*   *if the statement agrees with the information given in the passage;*

     N *(for **NO**)*   *if the statement contradicts the information given in the passage;*

     NG *(for **NOT GIVEN**)*   *if the information is not given in the passage.*

For questions 8~10, complete the sentence with the information given in the passage.

Rapid Police Response

A

Police departments in the United States and Canada see it as central to their role that responds to calls for helps as quickly as possible. This ability to react fast has been greatly improved with the aid of technology. The telephone and police radio, already long in use, assist greatly in the reduction of police response time. In more recent times there has been the introduction of the "9 · 11" emergency system, which allows the public easier and faster contact with police, and the use of police computer system, which assist police in planning patrols and assigning emergency requests to the police officers nearest to the scene of the emergency.

B

As an important part of police strategy, rapid police response is seen by police officers and the public alike as offering tremendous benefits. The more obvious ones are ability of police to apply first-aid life-saving techniques quickly and the greater likelihood of arresting people who may have participated in a crime. It aids in identifying those who witnessed an emergency or crime, as well as in collecting evidence. The overall reputation of a police department, too, is enhanced if rapid response is consistent, and this in itself promotes the prevention of crime. Needless to say, rapid response offers the public some degree of satisfaction in its police force.

C

While these may be the desired consequences of rapid police response, actual research has not shown it to be quite so beneficial. For example, it has been demonstrated that rapid response leads to a great likelihood of arrest only if responses are on the order of 1~2 minutes after a call is received by the police. When response times increase to 3~4 minutes—still quite a rapid response—the likelihood of an arrest is substantially reduced. Similarly, in identifying witnesses to emergencies or crimes, police are far more likely to be successful if they arrive at the scene no more than four minutes, on average, after receiving call for help. Yet both police officers and the public defined "rapid response" as responding up to 10~12 minutes after calling the police for help.

D

Should people police assume all the responsibility for ensuring a rapid response? Studies have shown that people tend to delay after an incident occurs before contacting the police. A crime victim may be injured and thus unable to call for help, for example, or no telephone may be available at the scene of the incident. Often, however, there is no such physical barrier to calling the police. Indeed, it is very common for crime victims to call their parents, their minister, or even their insurance company first. When the police are finally called in such case, the effectiveness of even the most rapid of responses is greatly diminished.

E

The effectiveness of rapid response also needs to be seen in light of the nature of the crime. For example, when someone rings the police after discovering their television set has been stolen from their home, there is little point, in terms of identifying those responsible for the crime, in ensuring a very rapid response. It is common in such burglary or theft cases that the victim discovers the crime hours, days, even weeks after it has occurred. When the victim is directly involved in the crime, however, as in the case of a robbery, rapid response, provided the victim was quickly able to contact the police, is more likely to be advantageous. Based on statistics comparing crimes that are discovered and those in which the victim is directly involved, Spelman and Brown (1981) suggest that three in four calls to police need not be met rapid response.

F

It becomes clear that the importance of response time in collecting evidence or catching criminals after a crime must be weighed against a variety of factors. Yet because police department officials assume the public strongly demands rapid response, they believe that every call to the police should be met with. Studies have shown, however, that while the public want quick response, more important is the information given by the police to the person asking for help. If a caller is told the police will arrive in five minutes but in fact it takes ten minutes or more, waiting the extra time can be extremely frustration. But if a caller is told he or she will have to wait 10 minutes and the police indeed arrive within that time, the caller is normally satisfied. Thus, rather than emphasizing rapid response, the focus of energies should be on establishing realistic expectations in the caller and making every attempt to meet them.

注意：此部分试题请在答题卡 1 上作答

1. Police believe there is a better chance of finding witnesses to a crime if response is rapid.
2. A response delay of 1～2 minutes may have substantial influence on whether or not a suspect criminal is caught.
3. The public and the police generally agree on the amount of time normally taken for a rapid response.
4. Physical barriers are the greatest cause of delay in contacting police.
5. Rapid response is considered desirable in handling cases of burglary.
6. Research shows that victims discover some 75% of crimes after they have been committed.
7. Police departments are usually successful in providing a rapid response regardless of the circumstances of the crime or emergency.

8～10 题请见答题卡 1

Part Ⅲ   Listening Comprehension   (35 minutes)

Section A

Directions: *In this section, you will hear 8 short and 2 long conversations. At the end of each conversation, one or more questions will be asked about what was said. Both the conversation and the question will be spoken only once. After each question there will be a pause. During the pause, you must read the four choices marked A). B), C) and D), and decide which is the best answer. Then mark the corresponding letter on the **Answer Sheet 2** with a single line through the center.*

注意：此部分试题请在答题卡 2 上作答

11. A) It was very small.   B) It was very expensive.
    C) It was very beautiful.   D) It was a fake.

12. A) The man thinks the match is wonderful.

    B) The man thinks the match is heated.

    C) The man doesn't think the match is wonderful at all.

    D) The man thinks the match is enjoyable.

13. A) In a department store.   B) In a laboratory.

    C) In a bookshop.   D) In a drug store.

14. A) He doesn't like the news.   B) He doesn't know it exactly.

    C) He doesn't agree with the woman.   D) He is ill.

15. A) Tommy will probably have the answer.

    B) They shouldn't tell Tommy about the mistake.

    C) It is Tommy who made the mistake.

    D) They don't think Tommy will take it.

16. A) Buy a new car.   B) Look for a less expensive car.

    C) Buy the car from the woman.   D) Help the woman paint her car.

17. A) He has already finished his report.   B) His report is already long enough.

    C) He will have time to finish his report.   D) He hasn't chosen a topic for his report.

18. A) She is tired of teaching.   B) She was dismissed from her job.

    C) She had a quarrel with the schoolmaster.   D) She moved to another city.

Questions 19 to 21 are based on the conversation you have just heard.

19. A) Whether they should go for a holiday.

    B) Where they should go for the holiday.

    C) How they could save enough money for the holiday.

    D) When they should leave for the holiday.

20. A) It is warmer than Sheffield.   B) It is too far away.

    C) It is colder than Sheffield.   D) They have been there.

21. A) Sheffield.   B) Hawaii.

    C) Wales or Scotland.   D) Sweden.

Questions 22 to 25 are based on the conversation you have just heard.

22. A) He can't find his office key.   B) He has misplaced some exams.

    C) He is unable to talk.   D) He doesn't like his classroom.

23. A) Mark the latest homework assignment.

    B) Put a cancellation notice on the classroom door.

    C) Make an appointment with the doctor.

    D) Return some exams to his students.

24. A) Teach Don's class while he's absent.

    B) Give Professor Webster the key to Don's office.

    C) Leave a message on the board in Don's classroom.

    D) Bring Don the homework that was due today.

25. A) To put the homework on Don's desk.

    B) To leave the master key for Don.

    C) To give Don's students the next assignment.

    D) To call Don at the end of the afternoon.

Section B

Directions: *In this section, you will hear 3 short passages. At the end of each passage, you will hear some questions. Both the passage and the questions will be spoken only once. After you hear a question, you must choose the best answer from the four choices marked A), B), C) and D). Then mark the corresponding letter on the **Answer Sheet 2** with a single line through the center.*

注意：此部分试题请在答题卡 2 上作答

Passage One

Question 26 to 28 are based on the passage you have just heard.

26. A) They are not well educated.

    B) They failed to work hard at school.

    C) The society is too complicated.

    D) What they learned in the school is adequate for their new life.

27. A) To find a worker and follow him closely.

    B) To find a person you admire and make friends with him.

    C) To find a person you respect and watch carefully how he acts.

    D) To make friends with a model you admire.

28. A) Learn From a Model   B) Learn, Learn and Learn Again

    C) Learn Forever   D) One Is Never Too Old to Learn

Passage Two

Questions 29 to 31 are based on the passage you have just heard.

29. A) He sold traffic-counting system to local governments.

    B) He dropped out of his law course at Harvard University.

    C) He founded his own company—Microsoft Software Company.

    D) He devised an operating system for IBM.

30. A) Windows.   B) A specialized software.

    C) MS-DOS.   D) Not mentioned.

31. A) At the age of 31.   B) At the age of 36.
    C) At the age of 20.   D) In 1986.

Passage Three

Questions 32 to 35 are based on the passage you have just heard.

32. A) Children who don't like to go to school.

    B) Children who are slow in study.

    C) Children who watch too much television during the day.

    D) Children who suffer problems from being left alone.

33. A) Give the keys back to their parents.   B) Hide the keys in their shirts.

    C) Leave the keys at home.   D) Cut down the number of their keys.

34. A) Tiredness.   B) Freedom.

    C) Loneliness.   D) Fear.

35. A) Latchkey children enjoy having such a large amount of time alone.

    B) Latchkey children try to hide their feeling.

    C) Latchkey children often watch TV with their parents.

    D) Some parents don't know the impact on children when they leave them alone.

Section C

Directions: *In this section, you will hear a passage three times. When the passage is read for the first time, you should listen carefully for its general idea. When the passage is read for the second time, you are required to fill in the blanks numbered from 36 to 43 with the exact words you have just heard. For blanks numbered from 44 to 46 you are required to fill in the missing information. You can either use the exact words you have just heard or write down the main points in your own words. Finally, when the passage is read for the third time, you should check what you have written.*

注意：此部分试题在**答题卡 2** 上，请在**答题卡 2** 上作答

Part IV   Reading Comprehension (Reading in Depth)   (25 minutes)

Section A

Directions： *In this section, there is a passage with ten blanks. You are required to select one word for each blank from a list of choices given in a word bank following the passage. Read the passage through carefully before making your choices. Each choice in the bank is identified by a letter. Please mark the corresponding letter for each item on Answer Sheet 2 with a single line through the centre. You may not use any of the words in the bank more than once.*

If you are the owner of an MBA that was suspiciously easy to earn, steer clear of Oregon. It is one of the few states in America that 47 pursues the sellers and users of fake degrees. Elsewhere, enforcement is largely up to employers, but the 48 can still be stiff: getting sacked, 49 your career and being exposed as a fraud. All across America, however, thousands of hard-working MBA evidently think that the risks of 50 are worth taking since in most places it is no straightforward matter for employers to sort the genuine qualifications from the fake one.

In credential-conscious America, thousands of such degrees are " 51 " every year. Some companies sell counterfeit diplomas from real schools, while others offer real degrees by bribing someone to insert fake files into a 52 school's systems.

A fake MBA—unlike a fake medical degree—will not kill anyone. But bringing a person with one can still prove expensive. Investing time and salary in someone who turns out to be 53 is bad enough, but if he makes an error before being caught, it could cost much more in lawsuits. So how to ensure that an MBA has been earned rather than 54 ? George Gollin, a professor at the University of Illinois, has posted a presentation describing how he tracked various schools he alleges are fake diploma mills. Putting the name of one of these schools into Google turns up a web-encyclopedia 55 of diploma mills. If a would-be employee has bought a degree from a diploma mill, 56 the name of its institution into a search engine will produce similar results.

注意：此部分试题请在答题卡 2 上作答

A) gradually	I) typing
B) rewarded	J) sending
C) bought	K) hurting
D) exposure	L) list
E) effective	M) ruining
F) unqualified	N) valid
G) awarded	O) penalties
H) aggressively	

Section B

Directions: *There are 2 passages in this section. Each passage is followed by some questions or unfinished statements. For each of them there are four choices marked A), B), C) and D). You should decide on the best choice and mark the corresponding letter on the **Answer Sheet 2** with a single line through the centre.*

Passage one

Questions 57 to 61 are based on the following passage.

What, besides children, connects mothers around the world and across the sea of time? It is chicken soup, one prominent American food, expert says.

From Russian villages to Africa and Asia, chicken soup has been the remedy for those weak in body and spirit. Mothers passed their knowledge on to ancient writers of Greece, China and Rome, and even the 12th century philosopher and physician Moses Maimonides *extolled* (赞美) its virtues.

Among the ancients, Aristotle thought poultry should stand in higher estimation than four-legged animals because the air is less dense than the earth. Chickens got another boost in the Book of Genesis, where it is written that birds and fish were created on the fifth day, a day before four-legged animals.

But according to Mimi Sheraton, who has spent much of the past three years exploring the world of chicken soup, much of the reason for chicken's real or imagined curative powers comes from its color.

Her new book, *The Whole World Loves Chicken Soup*, looks at the beloved and mysterious brew, with dozens of recipes from around the world. "Throughout the ages", she said, "there has been a lot of feeling that white-colored foods are easier to eat for the weak-women and the ill.

In addition, "soups, or anything for that matter eaten with a spoon" are considered "comfort foods", Sheraton said.

"I love soup and love making soup and as I was collecting recipes I began to see this as international dish—it has a universal mystique as something curative, a strength builder," Sheraton said from her New York home.

Her book treats the oldest remedy as if it was brand new.

The National Boiler Council, the trade group representing the chicken industry, reported that 51 percent of the people it surveyed said that they bought chicken because it was healthier, 50 percent said it was versatile, 41 percent said it was economical and 46 percent said it was low in fat.

注意： 此部分试题请在**答题卡 2** 上作答

57. Which of the following can be the best title of the passage?

   A) Prominent American Foods   B) Chicken Soup Recipes

   C) Chicken Soup, a Universal Cure-all   D) History of the Chicken Soup

58. Since ancient times, the value of chicken soup has been _____.

   A) widely acknowledged   B) over-estimated

   C) appreciated only by philosophers   D) has been known to mothers

59. Chicken soup earns universal praise because _____.

A) chicken soup has a very long history

B) chicken soup has curative power for those weak in body and spirit

C) poultry usually stands higher than four-legged animals

D) birds were said to be created earlier than four-legged animals

60. According to Sheraton, chicken soup has curative powers mainly for its _____.

A) taste   B) color   C) flavor   D) recipe

61. It can be said from the survey that chicken is _____.

A) a popular food   B) a main dish

C) cheaper than any other food   D) easy to cook

Passage Two

Questions 62 to 66 are based on the following passage.

Most episodes of absent-mindedness—forgetting where you left something or wondering why you just entered a room—are caused by a simple lack of attention, says Schacter. "You're supposed to remember something, but you haven't encoded it deeply."

Encoding, Schacter explains, is a special way of paying attention to an event that has a major impact on recalling it later. Failure to encode properly can create annoying situations. If you put your mobile phone in a pocket, for example, and don't pay attention to what you did because you're involved in a conversation, you'll probably forget that the phone is in the jacket now hanging in your wardrobe . "Your memory itself isn't failing you," says Schacter. "Rather, you didn't give your memory system the information it needed."

Lack of interest can also lead to absent-mindedness. "A man who can recite sports statistics from 30 years ago," says Zelinski, "may not remember to drop a letter in the mailbox." Women gave slightly better memories than men, possibly because they pay more attention to their environment, and memory relies on just that.

Visual cues can help prevent absent-mindedness, says Schacter. "But be sure the cue is clear and available," he cautions. If you want to remember to take a *medication* (药物) with lunch, put the pill bottle on the kitchen table—don't leave it in the medicine chest and write yourself a note that you keep in a pocket.

Another common episode of absent-mindedness: walking into a room and wondering why you're there. Most likely, you were thinking about something else. "Everyone does this from time to time," says Zelinski. The best thing to do is to return to where you were before entering the room. And you'll likely remember.

62. Why does the author think that encoding properly is very important?

    A) It helps us understand our memory system better.

    B) It enables us to recall something from our memory.

    C) It expands our memory capacity considerably.

    D) It slows down the process of losing our memory.

63. One possible reason why women have better memories than men is that _____ .

    A) they have a wider range of interests

    B) they are more reliant on the environment

    C) they have an unusual power of focusing their attention

    D) they are more interested in what's happening around them

64. A note in the pocket can hardly serve as a reminder because _____ .

    A) it will easily get lost

    B) it's not clear enough for you to read

    C) it's out of your sight

    D) it might get mixed up with other things

65. What do we learn from the last paragraph?

    A) If we focus our attention on one thing, we might forget another.

    B) Memory depends to a certain extent on the environment.

    C) Repetition helps improve our memory.

    D) If we keep forgetting things, we'd better return to where we were.

66. What is the passage mainly about?

    A) The process of gradual memory loss.   B) The causes of absent-mindedness.

    C) The impact of the environment on memory.   D) A way of encoding and recalling.

Part V   Cloze   (15 minutes)

Directions: *There are 20 blanks in the following passage. For each blank there are four choices marked A), B), C) and D) on the right side of the paper. You should choose the ONE that best fits into the passage. Then mark the corresponding letter on **Answer Sheet 2** with a single line through the center.*

注意：此部分试题请在答题卡 2 作答

For many people today, reading is no longer relaxation.
To keep up their work they must read letters, reports, trade
publications, interoffice communications, not to mention

newspapers and magazines: a never-ending flood of words. In __67__ a job or advancing in one, the ability to read

and comprehend __68__ can mean the difference between success and failure. Yet the unfortunate fact is that most of us are __69__ readers. Most of us develop poor reading __70__ at an early age, and never get over them. The main

deficiency __71__ in the actual stuff of language itself—words.

Taken individually, words have __72__ meaning until they are strung together into phrased, sentences and paragraphs. __73__, however, the untrained reader does not read groups of words. He laboriously reads one word at a time, often regressing to __74__ words or passages.

Regression, the tendency to look back over __75__ you have just read, is a common bad habit in reading. Another habit which __76__ down the speed of reading is vocalization—

sounding each word either orally or mentally as __77__ reads. To overcome these bad habits, some reading clinics use a device called an __78__, which moves a bar (or curtain) down the page at a predetermined speed. The bar is set at a slightly faster rate __79__ the reader finds comfortable, in order to "stretch" him. The accelerator forces the reader to read fast, __80__ word-by-word reading, regression and sub vocalization,

practically impossible. At first __81__ is sacrificed for speed. But when you learn to read ideas and concepts, you will not only read faster, __82__ your comprehension will improve.

Many people have found __83__ reading skill drastically

67.	A. applying	B. doing
	C. offering	D. getting
68.	A. quickly	B. easily
	C. roughly	D. decidedly
69.	A. good	B. curious
	C. poor	D. urgent
70.	A. training	B. habits
	C. situations	D. custom
71.	A. lies	B. combines
	C. touches	D. involves
72.	A. some	B. a lot
	C. little	D. dull
73.	A. Fortunately	B. In fact
	C. Logically	D. Unfortunately
74.	A. reuse	B. reread
	C. rewrite	D. recite
75.	A. what	B. which
	C. that	D. if
76.	A. scales	B. cuts
	C. slows	D. measures
77.	A. some one	B. one
	C. he	D. reader
78.	A. accelerator	B. actor
	C. amplifier	D. observer
79.	A. then	B. as
	C. beyond	D. than
80.	A. enabling	B. leading
	C. making	D. indicating
81.	A. meaning	B. comprehension
	C. gist	D. regression
82.	A. but	B. nor
	C. or	D. for
83.	A. our	B. your

improved after some training. __84__ Chalice Au, a business manager, for instance, his reading rate was a reasonably good 172 words a minute __85__ the training, now it is an excellent 1,378 words a minute. He is delighted that now he can __86__ a lot more reading material in a short period of time.

C. their   D. such a

84. A. Look at B. Take
  C. Make   D. Consider

85. A. for   B. in
  C. after   D. before

86. A. master B. go over
  C. present D. get through

Part VI   Translation   (5 minutes)

Directions: *Complete the sentences on Answer Sheet 2 by translating the Chinese given in brackets into English.*

注意：此部分试题在答题卡 2 上，请在答题卡 2 上作答

答题卡 1 （Answer Sheet 1）

Part I   **Writing**   **(30 minutes)**

Directions: *For this part, you are allowed thirty minutes to write a composition on the topic **Knowledge and Certificates**. You should write no less than 120 words and you should base your composition on the outline (given in Chinese) below.*

1. 生活中人们经常把知识和文凭联系在一起；
2. 然而非常有知识的人并不一定就要有文凭；
3. 我们也不应该想当然地认为有文凭的人就一定有知识。

Knowledge and Certificates

答题卡 1 （Answer Sheet 1）

_____
_____
_____
_____
_____
_____
_____
_____
_____
_____
_____
_____
_____
_____

Part Ⅱ   Reading Comprehension (Skimming and Scanning)   (15 minutes)

1. [Y] [N] [NG]   2. [Y] [N] [NG]   3. [Y] [N] [NG]   4. [Y] [N] [NG]

5. [Y] [N] [NG]   6. [Y] [N] [NG]   7. [Y] [N] [NG]

必须使用黑色签字笔书写，在答题区域内作答，超出以下矩形边框限定区域的答案无效。

8. The importance of response time in collecting evidence must be weighed against_____.

9. If a caller is told _____but in fact it takes ten minutes or more.

10. The focus of energies should be on _____ in the caller.

答题卡 2（Answer Sheet 2）

学校:

姓名:

填涂要求	正确填涂 ■
	错误填涂 ✓ ✗ ╱ ○ ● ▭

准 考 证 号

[0]	[0]	[0]	[0]	[0]	[0]	[0]	[0]	[0]	[0]	[0]	[0]	[0]	[0]	[0]
[1]	[1]	[1]	[1]	[1]	[1]	[1]	[1]	[1]	[1]	[1]	[1]	[1]	[1]	[1]
[2]	[2]	[2]	[2]	[2]	[2]	[2]	[2]	[2]	[2]	[2]	[2]	[2]	[2]	[2]
[3]	[3]	[3]	[3]	[3]	[3]	[3]	[3]	[3]	[3]	[3]	[3]	[3]	[3]	[3]
[4]	[4]	[4]	[4]	[4]	[4]	[4]	[4]	[4]	[4]	[4]	[4]	[4]	[4]	[4]
[5]	[5]	[5]	[5]	[5]	[5]	[5]	[5]	[5]	[5]	[5]	[5]	[5]	[5]	[5]
[6]	[6]	[6]	[6]	[6]	[6]	[6]	[6]	[6]	[6]	[6]	[6]	[6]	[6]	[6]
[7]	[7]	[7]	[7]	[7]	[7]	[7]	[7]	[7]	[7]	[7]	[7]	[7]	[7]	[7]
[8]	[8]	[8]	[8]	[8]	[8]	[8]	[8]	[8]	[8]	[8]	[8]	[8]	[8]	[8]
[9]	[9]	[9]	[9]	[9]	[9]	[9]	[9]	[9]	[9]	[9]	[9]	[9]	[9]	[9]

Part III

Section A			Section B	
11. [A] [B] [C] [D]	16. [A] [B] [C] [D]	21. [A] [B] [C] [D]	26. [A] [B] [C] [D]	31. [A] [B] [C] [D]
12. [A] [B] [C] [D]	17. [A] [B] [C] [D]	22. [A] [B] [C] [D]	27. [A] [B] [C] [D]	32. [A] [B] [C] [D]
13. [A] [B] [C] [D]	18. [A] [B] [C] [D]	23. [A] [B] [C] [D]	28. [A] [B] [C] [D]	33. [A] [B] [C] [D]
14. [A] [B] [C] [D]	19. [A] [B] [C] [D]	24. [A] [B] [C] [D]	29. [A] [B] [C] [D]	34. [A] [B] [C] [D]
15. [A] [B] [C] [D]	20. [A] [B] [C] [D]	25. [A] [B] [C] [D]	30. [A] [B] [C] [D]	35. [A] [B] [C] [D]

Part III   Section C

必须使用黑色签字笔书写，在答题区域内作答，超出以下矩形边框限定区域的答案无效。

If parents bring up a child with the(36)_____of turning the child into a (37) _____, they will cause a (38)_____. According to several leading (39)_____ psychologists, this is one of the biggest mistakes which ambitious parents make. Generally, the child will be only too (40)_____ of what the parent expects, and will fail. Unrealistic parental expectations can cause great (41)_____to children. However, if parents are not too unrealistic about what they expect their children to do, but are ambitious in a sensible way, the child may (42)_____ in doing very well — especially if the parents are very (43)_____ of their child.   Michael Li is very lucky. (44)_____.Although Michael's mother knows very little about music, Michael's father plays the trumpet in a large orchestra. However, he never makes Michael enter music competitions if he is unwilling. Michael's friend, Winston Chen, however, is not so lucky. (45) _____ _____.They want their son to be as successful as they are and so they enter him in every piano competition held. They are very unhappy when he does not win. (46) _____ Winston's father tells him. Winston is always afraid that he will disappoint his parents and now he always seems quiet and unhappy.

答题卡 2（Answer sheet 2）

Part IV	Section A	Section B	Part V
47. [A][B][C][D][E][F][G][H][I][J][K][L][M][N][O]	57. [A][B][C][D]	67.[A][B][C][D]	77. [A][B][C][D]
48. [A][B][C][D][E][F][G][H][I][J][K][L][M][N][O]	58. [A][B][C][D]	68.[A][B][C][D]	78. [A][B][C][D]
49. [A][B][C][D][E][F][G][H][I][J][K][L][M][N][O]	59. [A][B][C][D]	69.[A][B][C][D]	79. [A][B][C][D]
50. [A][B][C][D][E][F][G][H][I][J][K][L][M][N][O]	60. [A][B][C][D]	70. [A][B][C][D]	80. [A][B][C][D]
51. [A][B][C][D][E][F][G][H][I][J][K][L][M][N][O]	61. [A][B][C][D]	71. [A][B][C][D]	81. [A][B][C][D]
52. [A][B][C][D][E][F][G][H][I][J][K][L][M][N][O]	62. [A][B][C][D]	72. [A][B][C][D]	82. [A][B][C][D]
53. [A][B][C][D][E][F][G][H][I][J][K][L][M][N][O]	63. [A][B][C][D]	73. [A][B][C][D]	83. [A][B][C][D]
54. [A][B][C][D][E][F][G][H][I][J][K][L][M][N][O]	64. [A][B][C][D]	74. [A][B][C][D]	84. [A][B][C][D]
55. [A][B][C][D][E][F][G][H][I][J][K][L][M][N][O]	65. [A][B][C][D]	75. [A][B][C][D]	85. [A][B][C][D]
56. [A][B][C][D][E][F][G][H][I][J][K][L][M][N][O]	66. [A][B][C][D]	76. [A][B][C][D]	86. [A][B][C][D]

Part VI   Translation   (5 minutes)

必须使用黑色签字笔书写，在答题区域内作答，超出以下矩形边框限定区域的答案无效。

87. My mother would _____(大发脾气) if she found out about this.

88. Nobody knows how the universe _____(开始形成).

89. Furniture, books, clothes—_____(所有能够说得出来的) they sell it.

90. Let's start again from where we _____(停止).

91. What you said has_____(使人注意) to that matter.